# Topics in Classical Analysis and Applications in Honor of Daniel Waterman

# Topics in Classical Analysis and Applications in Honor of Daniel Waterman

Editors

Laura De Carli
Florida International University, USA

Kazaros Kazarian
Madrid Autonomous University, Spain

Mario Milman
Florida Atlantic University, USA

NEW JERSEY • LONDON • SINGAPORE • BEIJING • SHANGHAI • HONG KONG • TAIPEI • CHENNAI

*Published by*

World Scientific Publishing Co. Pte. Ltd.
5 Toh Tuck Link, Singapore 596224
*USA office:* 27 Warren Street, Suite 401-402, Hackensack, NJ 07601
*UK office:* 57 Shelton Street, Covent Garden, London WC2H 9HE

**Library of Congress Cataloging-in-Publication Data**
De Carli, Laura, 1962-
Topics in classical analysis and applications in honor of Daniel Waterman / Laura De Carli, Kazaros Kazarian and Mario Milman.
p. cm.
Includes bibliographical references and index.
ISBN-13: 978-981-283-443-0 (hardcover : alk. paper)
ISBN-10: 981-283-443-5 (hardcover : alk. paper)
1. Mathematical analysis. 2. Functional analysis. 3. Fourier series. 4. Orthogonal polynomials. I. Waterman, Daniel. II. Kazarian, Kazaros. III. Milman, Mario. IV. Title.
QA300.D346 2008
515--dc22

2008033323

**British Library Cataloguing-in-Publication Data**
A catalogue record for this book is available from the British Library.



Printed in Singapore.

## PREFACE

Some years ago, in order to try to re-energize the cultural life of analysts living in the tropics, Laura De Carli and Mario Milman developed the South Florida Analysis Seminar. They had the good fortune to count on Dan Waterman as one of the most enthusiastic participants.* So when they realized that Dan was going to turn 80, they naturally decided to devote the 2007 South Florida Analysis Seminar to celebrate the event in style. They enlisted the enthusiastic support of the chairpersons of the mathematics departments at FAU and FIU (Spyros Magliveras and Julian Edwards, respectively), who in turn enlisted the higher administrations of their schools. They also received the unconditional support of many of Dan's former colleagues, friends and former students and before very long, they had an impressive international conference under way.

The conference was held at FAU, Fort Lauderdale campus, on March 30–April 1, 2007. The main speakers at the Waterman feschrift were:

Marshall Ash (De Paul)
Calixto Calderón (Emeritus, UICC)
Michael Cwikel (Technion)
George Gasper (Emeritus, Northwestern)
Mourad Ismail (Central Florida)
Tadeusz Iwaniec (Syracuse)
Wolfgang Jurkat (Emeritus, Ulm)
Kazaros Kazarian (Autonoma de Madrid)
Togo Nishiura (Emeritus, Wayne)
Konstantin Oskolkov (South Carolina)
Eugene Poletsky (Syracuse)
Vladimir Temlyakov (South Carolina)
John Troutman (Emeritus, Syracuse)
Franciszek Prus-Wisniowski (Szczecin)

There were also many shorter talks by the *local* (and *non-local*!) participants of the regular SFAS.

The high quality of the conference, and the strong interest of the participants, led to the decision to publish a special volume. For this purpose, Laura and Mario decided to recruit Kazaros Kazarian to collaborate with them. All the participants were invited to contribute original research papers, and the manuscripts received

*And so there were three ...

were carefully peer-refereed. Professors Lardy and Troutman were invited to edit a special paper "Reminiscences", which contains a collection of testimonials contributed by some of Dan's friends and former colleagues and students. Professor Prus-Wisniowski, a former student of Dan, was invited to prepare a survey paper on recent progress on some problems dear to Dan. Finally, Dan Waterman himself graciously accepted an invitation to write an essay about his academic life. His paper includes some amusing stories, as well as a technical discussion of his contributions in different areas of classical analysis, and concludes with a complete list of his publications, with the names of all his former students.

We are very grateful to all the contributors to this volume, as well as to everyone that contributed in one way or another to the success of the conference.

Laura De Carli (Miami)
Kazaros Kazarian (Madrid)
Mario Milman (Delray Beach)

30th May 2008

## CONTENTS

# MY ACADEMIC LIFE

DANIEL WATERMAN

*Research Professor, Florida Atlantic University*
*(Professor Emeritus, Syracuse University), 7739 Majestic Palm Drive,*
*Boynton Beach, Florida 33437 USA*
*E-mail: dan.waterman@gmail.com*

*Keywords*: Reflexivity and summability, harmonic analysis, Fourier series and generalized variation, representation of functions, orthogonal series, real analysis.

## REMINISCENCES

I grew up under comfortable circumstances in Brooklyn, New York. I graduated from the local high school at the age of fifteen in 1943 and although I would have liked to be sent away to college, my parents believed that I was too young so I attended Brooklyn College. It took me some time to realize how fortunate I was. The New York City colleges had remarkable faculties due to the discrimination against Jews, Italians and women practiced at this country's most famous universities. The student body was also of high quality. Of the first eight Putnam examinations, Brooklyn won three, Toronto won four and Harvard won one.

My intended field of study was either medicine or dentistry. I was admitted to dental school while still a sophomore, but decided not to go, much to my mother's chagrin. Medicine seemed very appealing and a few of my teachers were suggesting that I study history or English or biology. At this point I registered in an advanced calculus course with Roger Johnson and it changed my life completely. In the first week we explored Dedekind cuts, Cauchy sequences and upper and lower limits. It was the first course I took in college that seemed like a real challenge and I gave up the thought of studying anything else. Of the dozen who finished that course, one went to Harvard, another to MIT and four, myself included, to Johns Hopkins. I graduated in the winter of 1947 and was asked to serve as an instructor in the spring semester to replace an ill faculty member. I recall vividly a moment when I was lecturing on trigonometry. A secretary came into the room with a telegram. I read it and turned to the class and said "It's just an offer from Harvard". The telegram was from Garrett Birkhoff Jr., offering me an assistantship. I rejected it and he called me to urge me to accept. I suppose he was not used to rejection. Nevertheless, I felt committed to Hopkins and I did not like the note of condescension I detected in Birkhoff's remarks.

The dominant figures in the Hopkins department were the chair, Francis Murnaghan, Aurel Wintner, and as a visitor, B. L. Van der Waerden. Van der Waerden was the best teacher I have ever had. His courses in topology and geometry were so exciting that the back of the classroom soon filled with faculty members and other auditors. In addition, he was remarkably friendly and helpful to the students. On the other hand, most of the students were fearful of Wintner and tried to stay out of his way. He was given to sudden changes of mood and could be most unpleasant. Hans Reiter came to study at Hopkins from Brazil, where his family had taken refuge from the Nazis in Austria. One day Wintner asked him a question in class, but Hans's English was so poor that he didn't understand. After class, he asked Wintner to please ask him questions in German. Wintner reacted by cutting off all contact with Hans. I had the opposite problem. My final examination paper in the first analysis course impressed Wintner greatly and I became his favorite. This meant that he started giving me books and problems to look at. For example, he suggested that I try to prove that the Mertens conjecture was false. I decided that I would leave after the first year. I had been looking at Antoni Zygmund's "Trigonometric Series" and applied to Chicago in the hope that he would take me as a student.

Chicago was very different. Marshall Stone was the chairman and he was very friendly and fair. He assumed that, since I had studied with Wintner, I must be very knowledgable about celestial mechanics. I did my best not to disillusion him. I greatly enjoyed the courses I took from him, Paul Halmos, Ed Spanier and Zygmund. The atmosphere was very informal in comparison with Hopkins, where the assistants were expected to wear a jacket and tie. On meeting Wintner in the hall, I always bowed slightly, exhibiting, to paraphrase W. S. Gilbert, the deference due to a person of high degree. One day I was standing in a corridor in Eckhart Hall with some of my colleagues, when Stone came striding along. I bowed and said "Good morning, Professor Stone". Once he had passed, my friends were almost hysterical with laughter and mimicked me for several days.

My association with Zygmund started only one year after he arrived in Chicago. I studied Fourier series, potential theory and $\mathcal{H}^p$ spaces with him. Later on, I graded papers for his course in measure and integral. At that time he was interested in trigonometric series and integrals in one and two variables, differentiation, harmonic functions, summability, and he also completed a set on notes entitled *Trigonometric Interpolation*, which became the basis for a chapter in the next edition of "*Trigonometric Series*". His work in singular integrals came toward the end of my stay there. Among my fellow Zygmund students were Berkovitz, Calderon, Cotlar, Shapiro and Wirszup. Other notable students at the time were Kadison, Singer, Michael, Bartle and Rosenberg. It was an exciting and stimulating environment.

It took me quite a while before I screwed up my courage and asked Zygmund if he would accept me as a doctoral student. I will never forget his response. He paused and looked at me very closely and said "Mr. Waterman, I have the impression that you are, how to say, somewhat lazy. If that is the case, you cannot work with me".

I assured him that this was not the case and he directed me to a recent paper of his on high indices theorems and suggested that I try to generalize his result. I was concerned about his opinion of me and hit on a plan to convince him that it was not correct. He was in the habit of going into the Mathematics Library several times a day. I would arrive early in the morning and start working at a table easily visible from the entrance and stay there most of the day. This seemed to work. The basic problem he gave me was to extend his result for $L^1$ to $L^p, p > 1$. Within a few weeks I could do it for $p = 2$ and then for rational $p$. I brought him the proof for $p = 2$. He wouldn't look at a handwritten proof, saying that I could give him a lecture on it or prepare a typewritten document. I gave him the lecture and indicated the stack of papers containing the argument for rational $p$. His comment was "There has to be a better way". I found the general argument and wrote it up within two months of my start. When I went to him with this I naively thought that this would be my dissertation, but within a few minutes he was describing another problem in a different area. When I had done that I indicated that I had thought I was done and he responded "Do you have children?" I told him I did not. He asked if I was married. Again the answer was no. He then said "Well, in that case I'm going to keep you around for a while and see what I can get out of you". Zygmund was gone for the quarter as I was finishing my dissertation research and Graves very kindly acted as my advisor, listening to me very patiently. In my last year and a half at Chicago, I was a research associate in the Cowles Commission for Research in Economics. Herstein was there at the same time and we became friends. The Commission was about to move to Yale at that time and I was invited to go with it, but I had received a Fulbright grant to the University of Vienna and I chose to go there in 1952. Although there were excellent mathematicians there, for example, Radon and Prachar, the mathematical climate was not particularly stimulating for a person of my interests. During the year Zygmund arranged two offers for me, and I chose to go to Purdue.

Lafayette and West Lafayette were sleepy little towns at that time, far different from anything I had ever experienced. The university hired twelve instructors that year, all with excellent credentials. We were a trial for the chairman; many of us did not come from places with the standards of dress and decorum that he was trying to maintain. However, adjustments were made on both sides. I made very good friends there, Michael Golomb, Lamberto Cesari, Casper Goffman and my fellow beginner, Robert Zink. Goffman and I became very close friends and collaborators and his influence altered my view of mathematics considerably. For a time, Paul Erdos was at Notre Dame and it was his habit to give up his hotel room on weekends, pack his belongings in two suitcases, leave one at the hotel and take the bus to Purdue, where he would stay with the Golombs. Whenever he got his paycheck, he would cash it and turn up at my place with a stack of five-dollar bills. He would have a list of charities to which he wished to contribute. We would sit together and I would write checks to these charities as he handed me the corresponding number of five-dollar bills. A substantial number of these contributions were to Native American groups.

It was from Paul that I learned of the plight of the people on the reservations, and to this day, most of my charitable contributions are to these groups.

My research blossomed in my time at Purdue. I wrote some substantial papers and produced two doctoral students. However I had a serious altercation with my dean. I taught an introductory real analysis course which was taken by undergraduates and a few graduate students. A new chair of an engineering department made the mistake of directing unqualified graduate students into the course. I was of the opinion that graduate students should be graded as undergraduates were. My dean wanted me to give graduate students higher grades than undergraduates received for similar achievement. I refused to change my grades. I was summoned to the dean's office where, in his Texas drawl, he told me "Dan'l, Purdue isn't big enough for the two of us. I guess you know what that means." It was like a scene from an old western movie.

I didn't look for another position. My friends at Purdue spoke to Morris Marden at the University of Wisconsin-Milwaukee and I was hired there. At that time, UWM was a tiny institution, recently formed by combining an extension of the University at Madison and a teachers college. It was just beginning a master's program. It turned out to be a very fortunate move for me. I met and married my wonderful wife Mudite there. In my second year there I received an offer of a full professorship at Wayne State University. Togo Nishiura, after completing his doctoral work with Cesari, had also come to UWM, and he also received an offer from Wayne.

Wayne was a stimulating environment for Nishiura and me. It had a strong group in analysis, including Vladimir Seidel, Frederick Bagemihl, Hidegoro Nakano, Albert Bharucha-Reid, Takashi Ito, and Leon Brown. The graduate students were also very strong. It seemed that Detroit had several gifted students who, for various reasons, were unable to leave the area. We also had some very good foreign students. In my years there, six students completed their doctoral dissertations with me.

Meanwhile, our family had grown; Mudite and I and our three lovely children went to Berkeley for a sabbatical during the 1967-68 academic year. On our return, we decided we needed more living space and purchased another house in Detroit. Before we were able to move, I received a very advantageous offer from Syracuse University. I accepted it, of course, and purchased another house in the Syracuse area. We now had two houses too many in the very depressed market which followed the Detroit riots of 1967. It took some ingenuity to dispose of them, but we did. Leaving Detroit was not so easy; we had made good friends there whom we would miss, particularly the Nishiuras. However Syracuse had many advantages including a superb library, an excellent environment for our children, and the presence of Wolfgang Jurkat, whose work I greatly admired.

Syracuse was an interesting institution. Don Kibbey, the chairman who hired me, was a person of great influence in the university and he used this influence for the benefit of the department. When he was forced to move upward in the administration, his power to help the department waned, and persons who had been jealous of his influence used his absence to deny the department many of the perquisites it had enjoyed. This made for many lean years. In addition, the main strength of the department had always been in analysis, but with the more democratic department structure which followed Kibbey's departure, some members of the other groups united to influence hiring tactics to the ultimate detriment of the department. As a frequent member of the department executive committee and ultimately the chairman, I learned much about the futility of making predictions concerning future research productivity based on early performance. Many gifted and productive people simply lose the drive which first inspired them. Perhaps in another department culture they would have fared better. Others fixate on one problem and may spend years on it without discernible progress. A strict department tenure policy would seem to be the solution, but often the faculty believes that the time allowed is too short and their personal feelings interfere with their scientific judgement.

My time at Syracuse was very satisfying. My research and my family thrived. My two daughters went to Cornell and Syracuse. They both became physicians. My son studied electrical and computer engineering at Berkeley and obtained a Ph.D. in computational linguistics at Brandeis. Mudite, who had a master's in mathematics from UWM, decided to study computer science. She reached the point of doctoral qualifiers, but decided that the children needed more attention than they would get if she continued, so she added another master's to her belt. I had eleven doctoral students at Syracuse and would have had more if I had not become chairman and if the stream of qualified students had not begun to dry up. I always found that supervising dissertation studies was enjoyable and also stimulated my own research. Being the chairman did give me some satisfaction. I was able to provide computer equipment and travel funds to the faculty that they might not otherwise have had, and with the help of my associate chair, John Troutman, was able to do some good things for our graduate students as well. During my term I made some notable appointments and, overall, I hired one quarter of the faculty.

I note that I had steady research support until I reached Syracuse. What happened then illustrates the tendency in this country for research support to follow fashion instead of relying on the abilities and judgement of the researcher. My first proposal to study generalized bounded variation met with a response on the order of "I have great respect for the previous work of the applicant, but I can't understand why he wants to do this". I was greatly gratified by the interest that was shown in my work by Eastern European mathematicians. It was widely cited and used. I had very pleasant letters from Orlicz and Chanturiya expressing their appreciation of this work. I am happy to see that the spaces of functions of generalized bounded variation I introduced are still the subject of study.

Retirement has had both its good and bad points. I have been graciously received by the mathematics department at Florida Atlantic University. I miss my friends in Syracuse and also its superb library. Modern computer technology and interlibrary loan can compensate to some extent, but nothing can replace the sensation of walking into a library with hundreds of the latest issues of mathematics journals arrayed on its shelves. I have managed to publish a paper per year since retirement and I also spend much time in communication with my former students and colleagues and in performing my editorial duties for the "Journal of Mathematical Analysis and Applications".

## RESEARCH

My current research interests are high-indices theorems, interpolating polynomials, and statistical summability. Statistical convergence of sequences was defined by Fast in 1950, and G. G. Lorentz and Sierpinski offered equivalent definitions independently. No suitable definition for continuous limits, e.g., for Abel means, has been given. I intend to pursue this question and try to find high-indices theorems for such methods of summability. In his text, Zygmund considered the partial sums of interpolating polynomials. I would like to estimate the degree of approximation of these partial sums to functions of various classes.

In my discussion I have grouped my work by area. Several papers could have been in more than one group. Papers will be referred to by the numbers in the publication list. References to papers of others will not be given; they are easily obtained from the cited papers. I cannot describe all of the papers in each group, but I will try to describe their main thrust as well as giving some extra attention to results that I am particularly fond of. Of course there are some papers that fall into none of the principal groups and I will not discuss these.

I can always recall the projects that I tried and failed to complete. One such project is very vivid in my memory. In the middle fifties I read a paper by W. Nef concerning regular functions on the quaternions. It was very interesting, but I discovered a crucial measure-theoretic error. I described this to my colleague, Artur Rosenthal, a renowned expert in real analysis, and he told me that it was impossible; he knew Nef and didn't believe that he could have done this. I showed him the paper and he confirmed that I was correct. I was strongly attracted by the idea. When I studied Hilbert space with Stone, we did it over the quaternions and I now envisioned doing harmonic analysis over the quaternions. My next thought was: why not do it over Clifford algebras? I wrote a detailed proposal and received a grant to pursue it. Unfortunately, my personal circumstances were such that I could not really undertake a project of this scope. I was able to pursue various problems with Goffman that were more limited and could be resolved in a shorter time, but maintaining the concentration necessary for this large project was impossible. I soon became captivated by the approach to Fourier series that we were pursuing and never returned to this project. These problems have since been taken up by many researchers and Clifford analysis has become a subject of considerable interest.

## High Indices

The high indices theorem of Hardy and Littlewood is a result about Abel summability, stating that if a power series $\sum a_k x^{n_k}$ has a finite limit as $x \to 1-$, where the sequence $\{n_k\}$ has Hadamard gaps, i.e., $n_{k+1}/n_k > q > 1$, then $\sum a_k$ converges. The theorem is valid for any sequence of real numbers $\{\lambda_k\}$ increasing in the same manner. It is convenient to set $x = e^{-s}$ and consider the limit as $s \to 0+$. We then have a Dirichlet series, $f(s) = \sum a_k e^{-\lambda_k s}$. Zygmund considered absolute Abel summability, showing that

$$\sum |a_k| \leqslant A_q \int_0^\infty |f'(s)|\, ds.$$

In [1], we showed that

$$\sum |a_k|^p \lambda_k^{p-1} \leqslant A_{qp} \int_0^\infty |f'(s)|^p\, ds,$$

for $p > 1$. Note that the Hardy Littlewood theorem and Zygmund's theorem are the extreme cases of this inequality, corresponding to $p = \infty$ and $p = 1$ respectively. Other results, with weights in the integrand, were also proved. We did not consider the case $0 < p < 1$.

Note that the integral in Zygmund's theorem is the variation of $f$. The interval of integration in both these results can be taken to be finite. This suggests that we consider the hypothesis that $f$ belongs to some class of functions of generalized bounded variation. We returned to this problem after fifty years and showed, in [72], that this result can be extended to $p \in (0,1)$ and that if we assume that $f \in \Gamma BV$ with $\Gamma = \{\gamma_k\}$, then

$$\sum |a_k| / \gamma_k \leqslant A_q V_\Gamma(f),$$

the variation extended over a finite interval $(0, B)$, whose length depends on $q$ and $\Gamma$. In [81], we establish a similar result for $f \in \Phi BV$.

In [12] we consider $f$, a gap series as above, with $s = \sigma + it$, the function being analytic in the right half-plane. Suppose $C$ is a curve terminating at the $s = 0$, on which $t \searrow 0$ as $\sigma \to 0+$. We give a Tauberian condition which ensures the convergence of the series at 0 if the limit of $f$ exists as $s \to 0$ along $C$.* [8] involves a similar limiting process.

## Reflexivity and Summability

The Banach-Saks theorem asserts that any bounded sequence in $L^p(0,1)$ or $l^p, p > 1$ has a subsequence whose $(C,1)$ means converge strongly. Banach spaces with this property are said to have the Banach-Saks property. Kakutani showed that

*We asserted that an analogous result would hold if $C$ terminated at another point on the vertical axis. The reviewer (*MR*) misunderstood my statement and said that this was incorrect.

for weakly convergent sequences in a uniformly convex Banach space the same conclusion holds. Since we now know that uniform convexity implies reflexivity, "weakly convergent" may be replaced by "bounded". A sequence-to-sequence summability method $T = (c_{mn})$ is regular if it satisfies the Toeplitz-Silverman conditions. A matrix satisfying the property $\sum_{n=1}^{\infty} |c_{mn}| \to 1$ as $m \to \infty$ is called *essentially positive.* A Banach space is said to have property $\mathcal{S}(w\mathcal{S})$ if for every bounded sequence there is a regular summability method $T$ and a subsequence whose $T$−means converge strongly(weakly). In [11], Nishiura and I showed that, for a Banach space $B$, the following three statements are equivalent: (i) $B$ is reflexive; (ii)[(iii)] $B$ has property $\mathcal{S}(w\mathcal{S})$ with essentially positive $T$. A. Baernstein has given an example of a reflexive Banach space which does not have the Banach-Saks property.

In [19], more general summability methods are considered. $T$ is *convergence preserving* if

(i) $\sum_{n=1}^{\infty} |c_{mn}| < H < \infty$ for every $m$;
(ii) $\sum_{n=1}^{\infty} c_{mn} \to c$ as $m \to \infty$;
(iii) $c_{mn} \to c_n$ as $m \to \infty$ for every $n$.

Here $c$ and $c_n$ are finite. $T$ is regular if and only if $c = 1$ and $c_n = 0$ for all $n$. A method is *almost regular** if it satisfies (ii), (iii), and

(iv) $c \neq \sum_{n=1}^{\infty} c_n$,

the latter sum being supposed convergent. When $c = 1$ and $c_n = 0$ for all $n$ an almost regular* method is *regular** or $T^*$ in the notation of Zygmund. We showed the following result:

*In a Banach space, property $w\mathcal{S}$ with almost regular* $T$ implies reflexivity, and reflexivity implies $\mathcal{S}$ with positive row-finite column-finite regular $T$.*

In [21] we discuss a paper of Klee in which he showed that certain Nakano spaces, $l^{p_i}$, contained bounded sequences with no $(C, 1)$ summable subsequences. We give necessary and sufficient conditions for the reflexivity of $l^{p_i}$, from which it is seen that the particular spaces he considered are not reflexive.

## Harmonic Analysis

In this classification I include papers on square functions and Fourier series on groups. Zygmund showed equivalence relations for the Littlewood-Paley functions $g$ and $g^*$. Thus if $f$ is a function in $L^p, p > 1$. on $(0, 2\pi)$, then for the corresponding $g$, we have

$$A_p \|g\|_p \leqslant \|f\|_p \leqslant B_p \|g\|_p ,$$

and similarly for $g^*$. He then showed such a theorem for the function $s(\theta)$, which is the square root of the area of the mapping by a function $f(z)$ in $\mathcal{H}^p, p > 1$, of a kite-shaped region terminating at $e^{i\theta}$. In [4] we proved similar results for functions analytic in a half-plane. He also considered the Marcinkiewicz function $\mu$ and proved a similar result. In [2, 5] we extended this to functions in $L^p(-\infty, \infty)$. The reviewer in MR said that this was done by "methods akin to those used by Zygmund", which is what I wrote about the proof of one of the inequalities. The proof of the other side was unexpectedly difficult. The referee of [5] asked me to shorten the paper and I complied. After it appeared, Zygmund told me that he was the referee and that he was sorry he had asked that, for it made the proof very difficult to follow.

In [15], we considered another problem related to area. This generalizes a result due to Lusin and Zygmund for the unit circle. Suppose $f(s) = \int_0^\infty e^{-sx} d\gamma(x)$, where $s = \sigma + i\tau$, is analytic in the half-plane $\sigma > 0$. Let

$$\alpha(x) = \sup_{0 \leqslant h \leqslant 1} |\gamma(x+h) - \gamma(x)| = o(1) \text{ as } x \to \infty.$$

Suppose $\Omega$ is a region in $\sigma > 0$ bounded by a segment $[i\alpha, i\beta]$ of $\sigma = 0$ and a Jordan arc. If $\iint_\Omega |f'|^2 \, d\sigma d\tau < \infty$, then $\int_0^\infty e^{-sx} d\gamma(x)$ converges a.e. on the segment $(i\alpha, i\beta)$ and uniformly on any closed subsegment of points of continuity. If $\alpha(x) = o(x^k)$, $k > 0$, we can replace convergence by $(C, k)$ summability. Only the argument for $(C, k)$ summability is given It involves dividing the Laplace integral into two parts and finding a trigonometric series such that it and its conjugate are uniformly $(C, k)$ equisummable with the two parts. This enables us to reduce the problem to that for the circle.

In [25, 29, 66] we consider functions defined on bounded 0−dimensional, metrizable, compact, abelian groups. Using the ordering defined by Vilenkin for the dual group, in [25] we generalize a result of Salem to Fourier series of continuous functions. This has several corollaries such as an analogue of the Dini-Lipschitz test. We also defined a notion of bounded fluctuation which is weaker than bounded variation. Functions satisfying this property had uniformly convergent Fourier series. In [29] we define a notion of harmonic bounded fluctuation, resembling harmonic bounded variation for trigonometric Fourier series. Continuous functions with this property are shown to have uniformly convergent Fourier series. An analogue of the Lebesgue test for continuous functions is proved. In [66], we proved a more general version of the Lebesgue test.

## Change of Variable

My interest in this subject began while working with Goffman on the convergence of the Fourier series of a function $f$ under every composition with a homeomorphism $g$. In [17] we found the condition on a continuous function which ensured this. The idea behind this was based on a linearization of the Dirichlet kernel used by Salem to prove a theorem on uniform convergence of Fourier series. In [28], a similar result is proved for preservation of uniform convergence and another proof of this is given in [55]. In [32] we showed that if a function $f$ was *equivalent* to a function $F$ that satisfied the condition of [17], then the Fourier series of $f \circ g$ would converge for every homeomorphism $g$. Equivalence here means that $f = F$ except on a set of *universal measure zero,* i.e., a set $E$ such that the Lebesgue measure of $g(E)$ is zero for every homeomorphism $g$ of $[-\pi, \pi]$ with itself. In this paper we assumed that the condition of [17] was also appropriate for functions whose only discontinuities were jumps, the regulated functions. Although this result can be demonstrated by methods similar to those for continuous functions, there are substantial difficulties, and the proof of this result appears in [70].

The result of [64] with Jurkat is one of which I am very fond. We showed that if $f$ is a continuous function on the circle group $T$, then there is a homeomorphism $g$ of $T$ onto itself such that the conjugate of $f \circ g$ is continuous and of bounded variation. This generalizes the Bohr-Pál theorem, which says that there is a continuous increasing $g$ mapping $[-\pi, \pi]$ onto itself such that the Fourier series of $f \circ g$ converges uniformly.

We define the Hadamard functions of *bounded deviation* to be integrable on $T$ and such that $\left|\widehat{f\chi_I(n)}\right| \leqslant C/n$ for a constant $C$, every integer $n$, and every subinterval $I$. In [31] we showed that $f \circ g$ is of bounded deviation for every homeomorphism $g$ if and only if $f$ is equivalent to a function of bounded variation. In [60] we refined this somewhat.

This result led naturally to consideration of the preservation of order of magnitude of Fourier coefficients under change of variable. Chanturiya defined the *modulus of variation* of a function $f$, $v(n, f) = \sup \sum_i^n |f(I_k)|$, $\{I_k\}_1^n$ running over all collections of disjoint subintervals. If $h(n)$ is a positive, nondecreasing, concave-downward function on the positive integers, then $V[h]$ is the class of regulated functions for which $v(n, f) = O(h(n))$. For regulated functions the following are equivalent: (i) $\left|\widehat{f \circ g(n)}\right| \leqslant C_f h(n)/n$ for every $g$; (ii) $\left|\widehat{f \circ g(n)}\right| \leqslant C_{f,g} h(n)/n$ for every $g$ (iii) $f \in V[h]$.

## Fourier Series and Generalized Variation

This begins with [27], where the notion of $\Lambda BV$ was introduced and applied to Fourier series. Let $\Lambda = \{\lambda_n\}$ be a nondecreasing sequence of positive numbers such that $\sum 1/\lambda_n$ diverges. If $f$ is a function defined on an interval such that $\sum |f(I_n)| / \lambda_n$ is bounded for all collections of nonoverlapping intervals $\{I_n\}$, $f$ is said to be of $\Lambda$ -bounded variation ($\Lambda BV$). If the $\lambda_n$ are bounded, we have the classical Jordan bounded variation ($BV$); if $\lambda_n = n$, we have *harmonic* bounded variation ($HBV$). It was shown that we can replace $BV$ by $HBV$ in the Dirichlet-Jordan theorem. If one were to use $\Lambda BV$ instead of $HBV$, where $\Lambda BV - HBV \neq \emptyset$, then the theorem would fail. Also, the relationship between the Banach indicatrix and $\Lambda BV$ was made clear. Paper [35] proved the properties of $\Lambda BV$ that were used in the previous paper and [39] gave a different proof of the generalized D-J theorem. Berezhnoi has shown that the generalization of the D-J theorem with $HBV$ is the strongest test for uniform convergence that can be obtained with generalized bounded variation.

The second major result in this area was the localization theorem with Goffman [36, 35]. We substituted $HBV$ for $BV$ in the Cesari-Tonelli definition of bounded variation for a function of two real variables and proved a localization theorem for double Fourier series. As described for the previous result, this theorem cannot be improved. When I presented this result to Zygmund, he responded enthusiastically, saying "This solves the localization problem for double series".

We have many other papers in which we explicate the properties of functions of generalized bounded variation, estimate the magnitude of Fourier coefficients, consider the degree of approximation of partial sums of Fourier series to functions of various classes, absolute convergence and many other topics which can be readily apprehended from the titles of the papers. There is, however, one to which I would like to draw attention.

We have previously alluded to a result of Salem which was a test for uniform convergence of Fourier series. Bary said of this that it appears "at first glance to be hardly suitable for application", but it has been the origin of many of our investigations; we refer, in particular to [61]. Let $f \in L^1(T)$ and, for odd positive integers let

$$T_n(x,t) = \sum_{k=0}^{(n-1)/2} \left[f(x+(t+2k\pi)/n) - f(x+(t+(2k+1)\pi)/n)\right]/(2k+1)$$

and let $Q_n(x,t)$ be obtained from this by replacing the $+$ sign after $x$ by a $-$ sign. These functions were defined by Salem. Our principal result was this: Let $x$ be a symmetric Lebesgue point of $f$, then the necessary and sufficient condition that the Fourier series of $f$ converge at $x$ is that

$$\int_{\pi}^{2\pi} (T_n(x,t) + Q_n(x,t)) \sin t dt = o(1) \quad as \quad n \to \infty.$$

Various definitions of bounded variation for functions of two variables have been proposed for over a century. Early definitions used 2-dimensional interval functions and later definitions, e.g., those of Cesari and Tonelli, employed variation over segments in one variable as a function of the other variable. Recently there have been results obtained by applying the notions of $\Lambda BV$ to intervals. The most successful have been the results of Dyachenko. Among these is our joint [79]. Suppose $f$ is defined on $A = [a,b] \times [c,d]$ and $\Lambda$ is as above. We say $f \in \Lambda^* BV$ if (i) $f(a,\cdot)$ and $f(\cdot,c)$ are in $\Lambda BV$, and (ii)

$$\sup_{\Gamma} \sum_k \lambda_k^{-1} \left|f(\alpha_k,\gamma_k) - f(\alpha_k,\delta_k) - f(\beta_k,\gamma_k) + f(\beta_k,\delta_k)\right| < \infty,$$

where $\Gamma$ is the set of finite collections of nonoverlapping subrectangles $[\alpha_k,\beta_k] \times [\gamma_k,\delta_k]$ of $A$. Various results were proved which relate this to previous definitions The principal result concerns continuity and convergence of Fourier series. We have

(i) If $f \in \Lambda^* BV$, then there exist at most countable sets $P \subset [a,b]$ and $Q \subset [c,d]$ such that $f$ is continuous at every $(x,y) \in A$ such that $x \notin P$ and $y \notin Q$ and, at every point of $A$, the limit of $f(x,y)$ from within each open quadrant exists;

(ii) If $f$ is $2\pi - \textit{periodic}$ in each variable and in $\Lambda^* BV(T^{2})$ with $\Lambda = \{n/\ln(n+1)\}$, then the rectangular partial sums of the Fourier series of $f$ are uniformly bounded and converge (Pringsheim) at each point to the arithmetic mean of the quadrant limits.

It is also shown that, in a certain sense, this convergence result cannot be improved.

Recent work has been in the area of trigonometric interpolation [80]. Zygmund studied the convergence of partial sums of interpolating polynomials and proved a theorem resembling the Dirichlet-Jordan theorem. We gave a different proof of this theorem, showed that bounded variation could be replaced by harmonic bounded variation, and that this result was best possible.

## Representation of Functions, Orthogonal Series

A result of Talalyan states that if $\{\phi_n\}$ is a basis of some $L^p$ space, where $1 \leq p < \infty$ then if any finite number of functions is deleted from $\{\phi_n\}$, the remaining sequence is still complete in the space of measurable functions $L^0$, with the topology of convergence in measure. This implies the existence of universal expansions and the existence of a subsequence $\{\phi_{n_k}\}$ which is complete in $L^0$ even though the complement of $\{n_k\}$ is infinite. We supplied a simple proof of a (more general) proposition where $\{\phi_n\}$ is only supposed to be complete in $L^0$ and the proof hinges on the fact that the dual of the space $L^0$ is trivial [7, 26]. With Kazarian, we made substantial generalizations of this result in [76].

Suppose $[\phi_n\}$ is a system of functions on (0.1) with $\phi_0 \equiv 1$. For $n = 2^{k_1} + 2^{k_2} + \cdots + 2^{k_s}, \{k_i\}$ an increasing sequence of nonnegative integers, set $\psi_n = \phi_{k_1} \cdot \phi_{k_2} \cdots \phi_{k_s}$. If $\{\psi_n\}$ is an orthogonal system on $(0, 1)$, it is called a *W-system.* It is generally assumed that $\int |\psi_n|^2$ is constant and $|\phi_n(x)| \leqslant 1$ a.e. for every $n$. Many results had been obtained that paralleled standard theorems on the Walsh system$\{w_n\}$, which is derived in this manner from the Rademacher functions. We showed [22, 49] that completeness of $\{\psi_n\}$ is equivalent to the existence of a $1 - 1$ measure-preserving mapping $\eta$ of $(0, 1)$ onto itself such that $\psi_n = w_n \circ \eta(x)$ a.e. for every $n$. Now let us consider a rearrangement of the sequence of Rademacher functions. If we rearrange the Walsh functions with indices $\{m_n\}$ obtained as described above, the summability and convergence behaviors of the series $\sum c_n w_n$ and $\sum c_n w_{m_n}$ are the same.

Shortly after this result was obtained I was reading the text of Alexits and found a problem of Steinhaus. I realized that an idea used in the proof of this result was relevant and obtained a very short solution [24]. Let $\{f_n\}$, $n = 1, 2, \ldots$be a sequence of measurable functions on $(0, 1)$. Consider the system $S = \{f_1^{m_1}(x) \cdot f_2^{m_2}(x) \cdots f_n^{m_n}(x)\}$, where $m_i = 0, 1, 2, \ldots, n = 1, 2, \ldots$. Steinhaus asked for the necessary and sufficient condition that $S$ be closed in $L^2$. Renyi introduced the following notion: $\{f_n\}$ is *maximal* if there is a set $Z$ of measure zero such that if $x_1$ and $x_2 \notin Z$ and $f_n(x_1) = f_n(x_2)$ for every $n$, then $x_1 = x_2$. He showed that maximality was sufficient. Now assume that $S$ is closed in $L^2$. Consider the function $f(x) = x$. There exists a sequence $\{P_n\}$ of linear combinations of elements of $\mathcal{S}$ such that $\|P_n - f\|_2 \to 0$ as $n \to \infty$. Thus there is a sequence $\{n_k\}$ such that $P_{n_k}(x) \to x$ a.e. in $(0,1)$ as $k \to \infty$. Let $Z = \{x : P_{n_k}(x) \nrightarrow x\}$. Then $m(Z) = 0$. Now if $x_1$ and $x_2 \notin Z$ and $f_n(x_1) = f_n(x_2)$ for every $n$, we have $x_1 = \lim P_{n_k}(x_1) = \lim P_{n_k}(x_2) = x_2$.

## Real Analysis

Much of what we have done in this area is connected to Fourier series and so may be found in that section. There are a few things that were not related to that area and are of some interest. The first paper I wrote with Goffman [6] arose from Menchoff's work on trigonometric series. He defined the concepts of upper and lower limits in measure for sequences of extended-valued measurable functions. His method involved transfinite induction, and the main properties of these limits were difficult to establish. By using the fact that these functions form a complete lattice, we were able to define the limits in a much simpler fashion and also to derive other expressions for them, e.g., for a sequence $\{f_n\}$, the upper limit in measure is $\inf[\overline{\lim}_{n\to\infty} g_n(x) : \{g_n\} \in \mathcal{F}]$ where $\mathcal{F}$ is the class of all $\{g_n\}$ such that $\{f_n - g_n\}$ converges in measure to zero.

Denjoy introduced the approximately continuous functions in his work on derivatives. In [9] we showed that the approximately continuous transformations from an Euclidean space into a metric space are of Baire class 1, have a Darboux property and have separable images. A measurable set in $E_n$ is *homogeneous* if it has metric density one at each of its points. $E_n$ can be topologized by taking the homogeneous sets as the open sets, yielding the *d-topology*. The approximately continuous functions are the continuous functions in the d-topology. We discussed connectedness and the generalization of the Darboux property. Although we discovered that Denjoy had implicitly defined this topology and it had been done explicitly in the text of Pauc, it had gone unnoticed and this paper led to a revival of interest in the subject.

## Summability

Suppose $\varphi$ is a nonnegative function on an interval to the right of the origin, that $\varphi(0) = 0$ and that $\varphi(t) = O(t)$ as $t \to 0$. A set $E$ is said to be $\varphi$-dense at a point $p$ if $m(E \setminus I)/\varphi(m(I)) \to 0$ as $m(I) \to 0$, $p$ in the interval $I$. If $\varphi(t) = t^\alpha$, $E$ is said to be $\alpha$-dense at $p$. The $\varphi\text{-}\lim_{ap\, t\to t_0} g(t) = a$ if, for every $\varepsilon > 0$, $\{t : |g(t) - a| < \varepsilon\}$ is $\varphi$-dense at $t_0$. Letting $\psi_{x_0}(h) = f(x_0 + h) - f(x_0 - h)$, we define the $\varphi$-approximate symmetric derivative at $x_0$ by $\varphi\text{-}f'_{aps}(x_0) = \varphi\text{-}\lim_{ap\, h\to 0} \psi_{x_0}(h)/2h$. A function $f$ satisfies condition $A_q$ at $x_0$ if, for some sufficiently large $M$, $\int_{E_M \cap (0,t)} |\psi_{x_0}(u)|\, du = o(t^q)$, where $E_M = \{t : |\psi_{x_0}(t)| \geqslant M\}$. In [16] we showed that if $f$ is integrable and $2\pi$-periodic, $\psi_{x_0}(h)$ is essentially bounded in a neighborhood of $t = 0$, and for $\alpha = 2$, $\alpha\text{-}f'_{aps}(x_0) = y$, then the differentiated Fourier series of $f$ is Abel summable to $y$

at $x_0$. In [20] we showed that essential boundedness can be replaced by condition $A_q$ with $q = 2$. This is best possible in the sense that $o(t^2)$ cannot be replaced by $O(t^2)$.

In [67] we showed that if $[f(x_0 + h) - f(x_0 - h)]/t$ is of harmonic bounded variation in a neighborhood of $t = 0$, then the differentiated Fourier series of $f$ is $(C, 1)$ summable to $\frac{1}{2}\left[f'_+(x_0) + f'_-(x_0)\right]$. A theorem on uniform summability is also proved.

In [71] we consider the Fourier series of functions in $\Lambda BV \supset HBV$. We assume $k/\lambda_k = O(1)$. A summability method $(W, \lambda)$ is defined which yields a D-J-like result for functions of this class with convergence replaced by $(W, \lambda)$ summability. A condition weaker than $\Lambda BV$ is defined which yields the same result, but this condition is not a generalization of bounded variation.

## Survey Papers

We wrote several survey papers, the first with Goffman on Fourier series [23]. We were very pleased that many people found this paper interesting and useful. I wrote several others on generalized bounded variation [37, 40, 41, 58] and [54] on change of variable. I regarded this as an important activity in that it often succeeded in bringing outstanding problems to the attention of others.

# PUBLICATIONS

## Papers

1. On some high indices theorems, Trans. Amer. Math. Soc. 69 (1950), 468–478.
2. On an integral of Marcinkiewicz, Proc. Int. Cong. of Math. 2 (1954), 185–186.
3. An extremal theorem for $n$-simplexes (with C. M. Petty), Monatsh. Math. (1955), 320–322.
4. On functions analytic in a half-plane, Trans. Amer. Math. Soc. 81 (1956), 167–194.
5. On an integral of Marcinkiewicz, Trans. Amer. Math. Soc. 91 (1959), 129–138.
6. On upper and lower limits in measure (with C. Goffman), Fund. Math. 48 (1960), 127–133.
7. Basic sequences in the space of measurable functions (with C. Goffman), Proc. Amer. Math. Soc. 11 (1960), 211–213.
8. A tangential Tauberian theorem, Monats. f. Math. 65 (1961), 101–105.
9. Approximately continuous transformations (with C. Goffman), Proc. Amer. Math. Soc. 12 (1961), 116–121.
10. Intersection of planar curves and straight lines (with T. Nishiura), Monats. f. Math. 67 (1963), 113–116.
11. Reflexivity and summability (with T. Nishiura), Studia Math. 23 (1963), 53–57.

12. A gap Tauberian theorem, Monats. f. Math. 67 (1963), 142–144.
13. Uniform convergence factors of orthogonal expansions (with S. A. Husain), Publ. de l'Inst. Math. Serbe 3 (1963), 89–92.
14. On the convergence of multiple power series (with T. Nishiura), Math. Zeitschr. 91 (1966), 277–279.
15. The local finite area principle in the half-plane, Proc. Amer. Math. Soc. 17 (1966), 1012–1015.
16. On the summability of the differentiated Fourier series, Bull. Amer. Math. Soc. 73 (1967), 109–112.
17. Functions whose Fourier series converge for every change of variable, (with C. Goffman), Proc. Amer. Math. Soc. 9 (1968), 80–86.
18. Functions whose compositions with homeomorphisms have everywhere convergent Fourier series (with C. Goffman), Orthogonal Expansions and Their Continuous Analogues (edited by D. T. Haimo), Carbondale and London 1968, 231–233.
19. Reflexivity and summability II, Studia. Math. 32 (1969), 61–63.
20. On the summability of the differentiated Fourier series II, Proc. Amer. Math. Soc. 20 (1969), 342–344.
21. Reflexivity and summability: The Nakano $l^{p_i}$ spaces (with T. Ito, F. Barber, and J.Ratti), Studia Math. 33 (1969), 141–146.
22. W-systems are the Walsh functions, Bull. Amer. Math. Soc. 75 (1969), 139–142.
23. Some aspects of Fourier series (with C. Goffman), Amer. Math. Monthly 77 (1970), 119–133.
24. On a problem of Steinhaus, Studia Scient. Math. Hungarica 5 (1970), 97–99.
25. Uniform convergence of Fourier series on groups I (with C. W. Onneweer), Mich. Math. J.18 (1971), 265–273.
26. A remark concerning universal series (with C. Goffman), J. Math. Anal. and Appl. 40 (1972), 735–737.
27. On convergence of Fourier series of functions of generalized bounded variation, Studia Math. 44 (1972), 107–117.
28. Functions whose Fourier series converge uniformly for every change of variable (with A. Baernstein), Indiana University Math. J. 22 (1972), 569–576.
29. Fourier series of functions of harmonic bounded fluctuation on groups (with C. W. Onneweer), J. d'Analyse Math. 27 (1974), 79–93.
30. On the invariance of certain classes of functions under composition (with M. Chaika), Proc. Amer. Math. Soc. 43 (1974) 345–348.
31. On functions of bounded deviation, Acta Scient. Math. Szeged 36 (1974), 259–263.
32. A characterization of the class of functions whose Fourier series converge for every change of variable (with C. Goffman), J. London Math. Soc. (2) 10 (1975), 69–74.
33. On the summability of Fourier series of functions of $\Lambda$- bounded variation, Studia Math. 5 (1976), 87–95.

34. The structure of regulated functions (with C. Goffman and G. Moran), Proc. Amer. Math. Soc. 57 (1976), 61–65.
35. On $\Lambda$-bounded variation, Studia Math. 57 (1976), 33–45.
36. On localization for double Fourier series (with C. Goffman), Proc. Natl. Acad. Sci. USA 75 (1978), 590–591.
37. Bounded variation and Fourier series, Real Anal. Exch. 3 (1977–1978), 61–85.
38. Some remarks on functions of $\Lambda$-bounded variation (with S. Perlman), Proc. Amer. Math. Soc. 74 (1979), 113–118.
39. Fourier series of functions of $\Lambda$-bounded variation, Proc. Amer. Math. Soc. 74 (1979), 119–123.
40. $\Lambda$-bounded variation: recent results and unsolved problems, Real Anal. Exch. 4 (1978-79), 69–75.
41. Generalized Bounded Variation-Recent Results and Open Problems, Real Anal. Exch. 5, (1978-79), 148–150.
42. Multiple Fourier series of functions of generalized bounded variation, Proc. of Symp. in Pure Math. 35 (1), (1979), 171–174.
43. Book review: Differentiation of real functions, by A. M. Bruckner, Bull. (N.S.) Amer. Math. Soc. 2 (1980), 232–237.
44. On the note of C. L. Belna, Proc. Amer. Soc., 80 (1980), 445–447.
45. The localization principle for double Fourier series (with C. Goffman), Studia Math. 69 (1980), 41–57.
46. A remark on the spaces $V^p_{\Lambda,\alpha}$ (with C. Goffman and F. C. Liu), Proc. Amer. Math. Soc. 82 (1981), 366–368.
47. On the summability of Fourier series (with B. N. Sahney), Rev. Roumaine de Math. 26 (1981), 327–330.
48. Estimating functions by partial sums of their Fourier series, J. Math. Anal. and Appl. 87 (1982), 51–57.
49. On systems of functions resembling the Walsh system, Mich. Math. J. 29 (1982), 83–87.
50. On the magnitude of Fourier coefficients (with M. Schramm), Proc. Amer. Math. Soc. 85 (1982), 407–410.
51. Absolute convergence of Fourier series of functions of $\Lambda BV(p)$ and $\Phi\Lambda BV$ (with M. Schramm), Acta Math. Acad. Sci. Hung. 40 (3-4) (1982), 273–276.
52. A differentiable function for which localization for double Fourier series fails (with C. Goffman and F. C. Liu), Real Anal. Exch. 8 (1982-83), 222–226.
53. On the rate of convergence of Fourier series of functions of generalized bounded variation (with R. Bojanic), Akad. Nauka Umjet. Bosne Hercegov. Rad. 74 (1983), 5–11.
54. New results on function classes invariant under change of variable, Real Analysis Exch. (1983-84), 146–153.
55. Functions whose Fourier series converge uniformly for every change of variable II, Indian J. Math. 25 (1983), 257–264.

56. Change-of-variable-invariant classes of functions and convergence of Fourier series, in Classical Real Analysis, Contemporary Mathematics AMS. 42 (1985), 203–211
57. Homeomorphisms of the circle and the magnitude of Fourier coefficients, Supp. Rend. Mat. del Circolo Palermo, Series 11, No. 8 (1985), 435–437.
58. Recent developments in Fourier analysis and generalized bounded variations, Real Analysis Exch. 12 (1986-87), 34–41.
59. On the preservation of the order of magnitude of Fourier coefficients under every change of variable, Analysis 6 (1986), 255–264.
60. On functions of bounded deviation II, J. Pure and Appl. Math., 131 (1988), 113–117.
61. A generalization of the Salem test, Proc. Amer. Math. Soc. 105 (1989), 129–133.
62. Fourier series with small gaps (with P. Isaza), J. Australian Math. Soc., (Series A) 46 (1989), 212–219.
63. Convergence of double Fourier series with coefficients of generalized bounded variation (with F. Moricz), J. of Math. Anal. and Appl., 140 (1989), 34–49.
64. Conjugate functions and the Bohr-Pal theorem (with W. Jurkat), Complex Variables, 12 (1989), 67–70. (Also appeared in abstract form, International Congress of Mathematicians, 1986.)
65. Uniform estimates of a trigonometric integral, Coll. Math. 60/61 (1990), 681–685.
66. On the Lebesgue test for the convergence of Vilenkin-Fourier series (with David Dezern), Mich. Math. J. 39 (1992), 425-434.
67. (C,1) Summability of the differentiated Fourier series (with P. N. Schembari), J. of Math. Anal. and Appl. 191 (1995), 633–646.
68. Smoothing $\Lambda$-sequences (with F. Prus-Wisniowski), Real Anal. Exch. 20 (1994-95), 647–650.
69. An integral mean value theorem, Real Anal. Exch. 21 (1995-96), 817–820.
70. Regulated functions whose Fourier series converge for every change of variable (with P. Pierce), J. Math. Anal. Appl. 214 (1997), 264–282
71. A summability method for Fourier series of functions of generalized bounded variation (with L. D'Antonio), Analysis 17 (1997), 287–299.
72. On some high-indices theorems II, J. London Math. Soc. 59 (1999), 978–986.
73. Bounded variation in the mean (with P. Pierce}, Proc. Amer. Math. Soc. 128 (2000), 2593–2596.
74. Necessary conditions for the convergence of Fourier series, Anal. Math. 26 (2000), 235–239.
75. A $\Delta_2$-equivalent condition (with P. Pierce), Real Anal. Exch. (2000-2001), 651–655.
76. Theorems on the representation of functions by series (with K. S. Kazarian), (Russian) Mat. Sb. 191(2000), 123–140; translation in Sb. Math. 191 (2000) 1873–1889.

77. On the invariance of classes $\Phi$-$BV$ and $\Lambda$-$BV$ under composition (with P. Pierce), Proc. Amer. Math. Soc. 13 (2004), 755–760.
78. On the magnitude of Fourier coefficients II (with M. Schramm), Analysis 24 (2004) 361–368.
79. Convergence of double Fourier series and W-classes (with M. I. Dyachenko), Trans. Amer. Math. Soc. 357 (2005), 397–407.
80. The convergence of partial sums of interpolating polynomials (with H. Xing), J. Math. Anal. Appl. 333 (2007), 543–555.
81. On some high-indices theorems III (with P. Pierce and M. Schramm), Analysis 28 (2008), 1001–1007.

## Books

1. Classical Real Analysis, (a volume of papers edited by D. Waterman), Contemporary Mathematics Series 42, Amer. Math. Soc., 1985.
2. Homeomorphisms in Analysis (with C. Goffman and T. Nishiura), Math. Surveys and Monographs vol. 54, Amer. Math. Soc. 1997.

## DOCTORAL STUDENTS

Syed Husain, Purdue University, 1959 - Orthogonal series
Dan Eustice, Purdue University, 1960 - Orthogonal series
Donald Solomon, Wayne State University, 1966 - Denjoy integral
Jogindar Ratti, Wayne State University, 1966 - Riesz summability
George Gasper, Jr. Wayne State University, 1967 - Harmonic analysis
James McLaughlin, Wayne State University, 1968 - Orthogonal series
Cornelius Onneweer, Wayne State University, 1969 - Fourier series on groups
Sanford Perlman, Wayne State University, 1972 - Harmonic Analysis
Elaine Cohen, Syracuse University, 1974, Fourier series
David Engles, Syracuse University, 1974. Generalized bounded variation
Arthur Shindhelm, Syracuse University, 1974 - Functional analysis
Michael Schramm, Syracuse University, 1982 - Generalized bounded variation
Pedro Isaza, Syracuse University, 1986 - Generalized bounded variation and Fourier series
Lawrence D'Antonio, Jr. Syracuse University, 1986 - Summability of Fourier series
David Dezern, Syracuse University,1988, Fourier series on groups
Nunzio Schembari, Syracuse University, 1991 - Fourier series
Pamela Pierce, Syracuse University, 1994, Change of variable and Fourier series
Franciszek Prus-Wisniowski, Syracuse University, 1994 - Generalized bounded variation
Hua-Ling Xing, Syracuse University, 1993 - Trigonometric interpolation

## REMINISCENCES

We were asked by the organizers of the Waterman conference on which this commemorative volume is based to collect reminiscences for inclusion. As colleagues of Dan's during his thirty years at Syracuse University, we were a reasonable choice, but it was a daunting assignment in view of Dan's many doctoral students (19 at last count), friends (too numerous to count) and colleagues. However, we contacted all those we could locate and received the responses given below. Several more wished to contribute but time constraints prevented their doing so.

Dan is a stimulating, open, generous and caring individual as the contributions here affirm.

Larry Lardy
John Troutman

I graduated from Syracuse University in 1986. Dan Waterman was my instructor in several courses and served as my Ph.D. thesis advisor. Dan has had a significant influence on my career, especially as a teacher. In fact, it's probably Dan's skill as a teacher that led me to ask him to be Ph.D. advisor.

My current research in the history of mathematics and bioinformatics may be far removed from the Fourier analysis that Dan and I worked on, but I still remember and value the classes that I took with Dan.

Let me convey what it was like to take a class with Dan. Typically we would be going through some proof when suddenly a problem would arise. Dan would get a puzzled look on his face: Why is this true? Or, is this really true? The normal classroom experience where the student passively listens to the professor has now broken down. We were all expected to help fill in the gap or correct the misbegotten proof.

I have never known whether these interruptions where intentional or accidental. In any case they showed me an important lesson. Mathematics is not a performance art; it is a struggle to attain truth. The students have to be active participants in this struggle.

One particular day comes to mind. Dan had some problem standing in class (perhaps it was his foot). So he used an overhead projector. We were merrily rolling along through the proof of that day's theorem. Suddenly that puzzled look comes on Dan's face. This time we were all seriously stuck. Dan came over to sit with us in order to look up at the projection screen. At that time I happened to notice an undergraduate student walk by the classroom. I saw him look in upon us and then stop in his tracks with jaw agape. From his perspective he saw a classroom with no apparent teacher, with everyone in class silently staring into space. I'm sure that he never saw such a sight!

I'm sure that what was most surprising to that student was the sight of a class actually thinking. For this and many other valuable experiences I want to deeply thank Dan Waterman.

Lawrence D'Antonio

Professor Waterman's friends sometimes referred to him as "Doctor Dan," not because he had earned a doctorate in mathematics, but because of his extensive knowledge of medicine. Time has dimmed my memory, but as I recall his mother-in-law was put on the powerful corticosteroid drug prednisone because she was suffering from temporal arteritis. For some reason her physician took her off the medication precipitously. Since Dan knew that this was a very dangerous action, he made sure that she was put back on the medication and then taken off it gradually. On another occasion, when Dan learned that my wife had started taking a certain prescription drug, he suggested that she also take the dietary supplement CoQ 10, he told her the physiological reason why she should do so, and he told her the best place to procure the supplement.

When Dan was Chair of the Department of Mathematics, he was genuinely concerned about the personal problems of his faculty and staff. Once, when a staff member was experiencing a personal crisis, Dan took up a collection for the staff member.

Although Dan's mathematical training under the harmonic analyst Antoni Zygmund was classical, Dan was well versed in abstract mathematics. On one occasion when he was casually discussing preliminary examinations with a graduate student, Dan started talking about a rather abstract theorem in functional analysis that would be foreign to most classical analysts trained in the 1950s.

Gerald T. Cargo

— ◊♦◊ —

Dan Waterman was a very fine colleague for his many years at Syracuse University, especially during the years 1988 to 1994 of his chairmanship. Although our research interests were disjoint, we worked together over these years on many delicate personnel matters and I found him to exercise very thoughtful judgment.

In the spring of 1971 our long time Department Chair, Donald E. Kibbey, was promoted to Vice-President for Research. The department's faculty elected Erik Hemmingsen chair, voted that there should be a department constitution, and elected four individuals to draft it: Tekla Lewin, Dan Waterman, the late David Williams and me.

To a large extent the document consisted of a careful description of existing practice, but there was one major change: a careful setting up of procedures for faculty personnel matters, viz. renewal and non-renewal of appointment, tenure and promotion. This was Dan Waterman's suggestion, and in retrospect was a prescient move. Not so long afterward the move nationally was toward more procedural safeguards and lawsuits when universities were not careful.

Philip T. Church

My first encounter with Dan Waterman came in the introductory real analysis course at Syracuse. I was one of those students who sat in the middle of the classroom — not in the front, not in the very back — and during that first semester I probably said no more than a couple of words in class. I kept a low profile because I found the instructor intimidating. I needn't have worried. A few years later, when I finally followed Mike Schramm's advice to ask Waterman to become my dissertation advisor, I could not have found a warmer, more supportive and encouraging mentor. He took a real personal interest in me — once picking me up when my car ran out of gas — and I came to value his well-informed opinions on everything. Some bits of advice he gave me were more helpful than others. (Does anyone remember the cabbage soup diet?) But he directed me to an honest auto mechanic, a great optometrist, and I would have gone to his dentist too, except that! I didn't want to travel all the way to Rochester to get my teeth cleaned.

And of course he taught me about Fourier series. I have very fond memories of walking up to his office in the physics building, where there would always be hot water for tea (sweetened with that banned artificial sweetener he had brought back from Brazil), and telling him about my progress on my dissertation. One day I came in with a little lemma over which he shook his head and said, "No, this can't be right." So I went away and worked through the computation again and convinced myself that it really was correct. Our next meeting, however, produced the same outcome: I would leave the office sure that there had to be a mistake lurking somewhere in what I had written, but after poring over it I would again come to the conclusion that the computation was right. Altogether it took several sessions over about two weeks to convince him of this, and then he made me change the presentation of the lemma so that it didn't look so strange — so that actually the end result bore very little resemblance to the original. I think back on this episode as one of the high points of my academic career, with the back and forth and give and take of our discussions comprising one of the most delightful intellectual experiences I have ever had.

Having been out of touch with Dan and my contemporaries for some time (I worked outside academia for about ten years and returned to teaching only in 2005), I was very happy to rejoin the group at the conference in honor of his birthday. I thank the editors of this volume for this opportunity to express to Dan my admiration and my gratitude.

David Dezern

One of my earliest memories of Dan Waterman is the way he taught his "Fourier Analysis" course at Wayne State University. In his class many mathematically stimulating hours were devoted to meticulously filling in the not-so-obvious missing details in the proofs of some of the main theorems in Zygmund's "Trigonometric Series" book. This laid a foundation that enabled the students to derive new theorems. It was during the summer of 1966, while reading some of Dan's reprints, that I discovered a new method for deriving the $p$-norm inequalities for the Littlewood-Paley and Lusin functions in Dan's 1956 "On functions analytic in a half-plane" paper. The $n$-dimensional extension of this method enabled me to extend Dan's norm inequalities to systems of conjugate harmonic functions in both the $n$-dimensional unit ball and Euclidean half-space. Dan encouraged me to quickly write a short research announcement entitled "On the Littlewood-Paley and Lusin functions in higher dimensions" to send to his thesis advisor, A. Zygmund, who kindly communicated it for publication in the January 1967, Proceedings of the National Academy of Sciences. Another vivid memory concerns an April, 1967, car trip in my compact Ford Falcon with Dan and a couple of his graduate students from Detroit, MI, to Edwardsville, IL, to attend a conference on "Orthogonal Expansions and their Continuous Analogues" at Southern Illinois University. After sharing the driving Dan volunteered to sit in the small rear seat. When I expressed my concern about the rear seat being too uncomfortable for him, he responded that the seat's stiff vertical back was great for his back. Among the many papers presented at the conference was the Askey-Wainger "A dual convolution structure for Jacobi polynomials" paper.

In a 1970 paper I was able to determine for which $(\alpha, \beta)$ all of the coefficients in the linearization for the product of Jacobi polynomials were nonnegative, and thus extend the results in the Askey-Wainger paper. Also, G. Szegö presented his penultimate paper "An outline of the history of orthogonal polynomials" at that conference.

In a 1972 paper I was able to generalize the inequality in Szegö's 1962 "An inequality for Jacobi polynomials" paper by determining all $(\alpha, \beta)$ such that the Turán inequality holds for Jacobi polynomials. Dan was always readily available to his graduate students to discuss their research projects with them. He was certainly a major influence in shaping my early mathematical career. I am indebted to Dan for his guidance and encouragement.

George Gasper

Dan Waterman has had an impressive career in mathematics, and he is an outstanding mentor and role model for many aspiring mathematicians. His papers are too numerous to mention; giving an exact number would not do justice to the quality and scope of his work. Dan has published on a wide variety of subjects and his papers have influenced research in real analysis all over the globe. He has defined and studied the field of $\Lambda$-bounded variation, and this in turn has opened the doors to further research by numerous colleagues and graduate students. I am fortunate to be among those graduate students who have had the pleasure of working under Dan's guidance. Early in my graduate school career, Dan gave a colloquium talk that grabbed my attention, and I knew right then that it would be fun to work with Dan. His eyes lit up as he sketched some "wild" functions on the board, and I could tell that he enjoyed studying the properties of functions on the real line just as I did. This was someone who thought deeply about mathematics, and I found his keen interest in the subject to be inspirational. My meetings with Dan were never dull. The math itself, though difficult, was engaging, and we found time to talk about other things. It was during these meetings that I discovered that Dan knows something about everything. He gave me advice on automobiles, driving, dentists, medical remedies, how to get along with others, and how to eat a piece of fruit. The door to his office read "Daniel Waterman, Department Chair", but it should have read "Mathematics Help: free, Psychiatric Help: 5 cents, Other Pearls of Wisdom: priceless". There were many graduate students who would not have made it through their program successfully were it not for Dan's wise counsel and encouragement.

Dan was not afraid to tell me to work harder, to budget my time more effectively, and to think more about mathematical problems. Dan will tell people the truth, even if it hurts, because he wants to see them be the best that they can be. His advice always comes from the heart, showing genuine care and compassion for his friends. Fortunately for me, Dan did not see the mentoring relationship between us as one that would end after the thesis defense-quite the contrary. Dan has continued to mentor me in several important ways. He has helped me in my career path by continuing to work with me as a co-author on several papers, and he has encouraged me to branch out and work with others as well. Without his hard work and clever insight, a recent paper that I was working on with Mike Schramm would still be on my desk, instead of in the hands of a journal editor. I am grateful to Dan for being my mentor, because of the learning that I have been able to accomplish with his guidance, and because along the way he became my friend. In ways that go well beyond his mathematical insights, there is no doubt that he has played a huge role in my life.

Pam Pierce

In May of 1989 my wife and I landed in JFK to start the new life in US after 10 years in waiting. Two weeks later our friends Tadeusz and Grazina Iwaniec drove us to their house in Syracuse. During our refusniks period I became a rather accomplished programmer but my dream was an academic career in mathematics. But my CV was not impressive enough to generate a flow of offers.

Fortunately, the Department of Mathematics of Syracuse University had a vacant visiting position and Tadeusz convinced Dan that I could fill it for one year. This was the beginning of what I consider now to be the best 19 years of my life (excluding the first 19, of course). It was not a paved road: I failed to get an offer next year and Dan kept me as a visitor for another year. In a year, I believe, he made great efforts to convince the executive committee to give me a tenure-track position. At that time the Great Wall separating the East from the West had crumbled and huge hordes of eastern mathematicians of exceptional talents were filling academic positions in US. Since my age and my achievements were not in harmony, mathematics did not seem to be an option.

Looking back, I still cannot understand why Dan believed in me. But, in my, perhaps, exaggerated, opinion he should not be ashamed of his trust. And I am infinitely grateful to him for making my life what it is. Probably, my programming abilities could have brought me a good job in a software company and, who knows, even millions of dollars. But loser's worm would forever eat my heart. Thank you, Dan!

My mathematical interaction with Dan was mostly in real analysis. Doubtfully, he learned something from me. Among my real analysis interests were Suslin sets, measurable selections and similar stuff coming from complex analysis. But his influence on me was very important, at least, in one instance. Finishing in 1991 my magnum opus, which brought me international recognition, I decided to give its 40 pages to Dan for comments. My expectations were not too high: long text, unfamiliar subject. Returning it to me rather soon Dan asked: "Why do you call some functions almost continuous?" I tried to explain the notion but he continued: "Everybody calls them approximately continuous." As it happened the additional information, obtained from Dan, about this class of functions was rather influential for my consequent studies. My most beloved subharmonic and plurisubharmonic functions are approximately continuous and in many questions the condition of upper semicontinuity imposed on them is too restrictive. At these moments the notion of approximate continuity was very helpful and, from time to time, I think that in a good potential theory the latter notion should be used instead of the classical upper semicontinuity.

Another thing that I learned with Dan's help was nonstandard analysis. He was excited by its applications to many standard questions and organized a seminar devoted to this subject. Although it was never used in my research, the acquaintance with this strange world expanded my mathematical vision.

This experience is very characteristic of Dan. He is permanently interested in mathematics and always tries to learn more of it. Several years ago he said to me: "All my life I dreamed of reading the proof of the Fermat Theorem. And now Wiles found it and it is too algebraic." The sadness with which it was said shamed me. In no way would I be able to penetrate Wiles's proof which requires a vast amount of knowledge, but that will not influence my mood. After that it became my persistent wish to find a simpler proof especially for Dan.

But what I liked most about Dan is his art of partying. Being the chairman of the department he was responsible for the annual faculty gatherings. His method was simple and effective: rent a barn, invite caterers with chicken drumsticks and hamburgers, buy, let's say, an inexpensive wine from Romania and let people mingle freely. As a result, with enough food in our stomachs and wine in our hearts, we would enjoy several hours of fun.

Later, my wife and I acquired the habit of spending the winter breaks in Florida, where Dan and Mudite migrated after retirement. They lived not far from our usual place of enjoyment and every New Year's Eve we met in their house. On every such occasion the cooking was Dan's responsibility. His devotion to this process and his level of confidence were immense. Once he said to us: "The dish for tonight was cooked for you by me for 12 hours." As it happened, he had bought a slow cooker.

Tatyana and I always enjoyed the possibility to spend these hours, when you should be awake in expectation of the completion of Earth's revolution, in the warm environment of Waterman's house heated by an amicable conversation and a generous amount of single malt whiskey that was always in abundance.

Concluding these reminiscences I want to tell about Dan's latest visit to Syracuse. Tatyana organized a nice party for him, Mudite and other of Dan's friends in our house. I convinced Dan to give a colloquium talk in the department so the food and wine in the party would be well earned. He agreed and it was a real fest. He talked about the summation of Fourier series, a topic that is not my favorite, but the quality of results and the mathematical taste in their statements were incredibly impressive. Here was a man who in ten years of his retirement published 12 papers of impeccable taste; and the only thing you could do was to bow.

Evgeny Poletsky

I was at Syracuse for a few years before I met Dan, who was viewed by graduate students with an air of mystery, tinged possibly with a little dread.

There were rumors of homework questions that didn't have answers in the usual sense, never knowing how seriously to take his suggestions that a few weeks of study might be necessary to understand a single question, and the terra incognita of taking a course where the instructor was still actively interested in the material.

But, though I certainly knew of him, I had never actually met Dan or taken a course from him until after Dave Williams' death. I had begun working with Dave, and though I was not too far along, I had done enough to be, among all the other emotions of that time, panicked at the loss of a mentor. Dan approached me and assured me that things would work out.

This, in a nutshell, is what Dan Waterman means to me: With what may well have been the first words we had ever spoken, he offered to help.

And it hasn't changed since then. The week after our first chat, I found a pile of papers on my desk, delivered with the comment "see if you can improve on these." This seemed like a daunting task, since the papers were mostly Dan's. But Dan's brilliance as a thesis advisor always was his ability to pick just the right topics for each of his students – mysterious enough to be captivating, but within reach given the effort of which Dan somehow knew you were capable.

Dan has been gracious to ask me to work on some of his papers with him, which I find to be a great honor and privilege. I consider each of these opportunities to be another gift, which I'm hard-pressed to return in kind.

Dan is one of the most caring, giving, and worldly people I have ever met. Aside from the mathematics, he taught me that this is a good way to go through life.

Mike Schramm

— ◊♦◊ —

My acquaintance with Dan Waterman started through his math writings in the Real Analysis Exchange. In the year 1987, I was telling one of my colleagues (a differential geometer) how easy, comparing to his field of research, are concepts of generalized bounded variation and their applications. I happened to have a copy of RAE on my office desk, so I took it to write the definition of a function of bounded $\Lambda$-variation on the blackboard, and next I read aloud some of the open questions in this area of math. Of course, they were questions collected by Dan in one of his survey articles on the topic. My colleague agreed with me that the questions are elementary-like and we discussed the problem of relationship between ordered $\Lambda$-variation and the plain one for more than two hours - seemingly to no avail. However, I continued to think about it at home most of that afternoon, and I got a partial answer somewhat better than the theorem of Belna. A few weeks later I solved the problem completely (during a three-hour-long train trip to Poznan, to Professor Musielak's seminar on modular spaces). Although my main field of research was functions of bounded $\Lambda$-variation - with the intention to write a Ph.D. thesis about them under the supervision of our local math department head - I kept thinking about other open questions posed by the unknown guy called Dan Waterman. This led me to prove a decomposition theorem on local $\Lambda$-variation which eventually yielded a characterization of functions continuous in $\Lambda$-variation, answering another of Dan's open problems. I submitted the result to Studia Mathematica, where the relevant question was posed in 1976 for the first time. The journal editors decided not to publish my article despite a positive referee's report. I got angry at them, and then it occurred to me to send my writings on $\Lambda BV$ to the author of the concept. In the spring of 1989 I sent two or three manuscripts to Dan Waterman, and a little later I received a friendly reply, praising my math and inviting me to consider cooperation, possibly by coming to Syracuse University as a visiting professor. When I explained that I had not yet gotten my Ph.D., in his next letter in July 1989, Dan offered me a teaching assistantship at Syracuse University (starting August 1989). I had to explain that despite being delighted at the prospect of studying in the USA under his guidance I was forced to decline the offer because of my ignorance of spoken English. I promised that if he could wait one year I would learn the language well enough to pass the necessary TOEFL exam. My wife (also a beginning mathematician) took most of household duties on herself (she gave birth to our second daughter in September 1989) and that sacrifice made it feasible for me to fulfill my promise to Dan. He kept his word and I received a 4-year Syracuse University Teaching Fellowship. On August 14, 1990, I met Dan in his office in the Carnegie Hall. I took a picture of him then. I still have it, but I don't need it. I have it in my mind forever.

A few days later, I told Dan that my trip to the US had separated me from the rest of my family, and that I had not only a wife, but also two daughters. I wanted to return to them after one year. "Don't worry about that", he said, "We'll find a solution". Indeed, with Dan at my side I did not have to be worried. Ewa applied for a TA position at Syracuse University's Math Department and got it swiftly. Dan made sure that we could study together at SU and have our kids with us. He helped us one year later when he saw that it was too difficult for Ewa to be a grad student in pure math and a mother of two at the same time. Dan suggested that Ewa switch to the math education Ph.D. program and write her thesis under the supervision of Professor Joanna Masingila. Ewa followed Dan's advice. It was a perfect solution. Ewa found an enjoyable and interestiong area to study, and finished her Ph.D. in Math. Education in 1995.

At first Dan and I didn't talk much about math. He was busy running the department, and I was busy learning English and preparing for Qualifying Exams. He turned my attention to the unsolved problem of the relationship between the Garsia-Sawyer class and HBV, but my vague attempts to solve it failed. Dan was taking care of me and my family, but he did it very subtly. For instance, every year we got summer courses to teach — a crucial thing for supporting our family during months with no assistantship grant. Of course, there were always fewer summer courses than TA's willing to teach them.

Later on Dan introduced me to the world of attending math conferences. He was able to dig up funding for me to go to various real analysis meetings about twice a year. The mathematics at the meetings was good, people were very friendly and I could relate myself to the math community. On some occasions Dan dined with me at curious ethnic restaurants, teaching me the pleasure of enjoying a variety of new tastes. Some of them have remained with me until today, including the bottle of retsina we tried in a quiet restaurant in Baltimore. (Dan did not want to go to any of the fancy places in the center of the city. "Their cuisine is not ethnic, it's ethnic-like", he explained. We took a cab and Dan asked the driver if he could suggest a good restaurant, preferably Thai or Chinese (such were our plans). The driver knew only one restaurant he could recommend. He took us far away to a Greek restaurant in the Greek district of the city. The place look moderately inviting, but the food was excellent. (Even Dan was impressed.)

One day at Syracuse Dan invited me into his office to discuss some math with me. At first he spent a few minutes trying to clear out some space on his large desk covered completely with enormous stacks of papers and books. He did not succeed, so he moved to his even larger oval table covered with even higher piles of papers. After a while Dan managed to remove some of them, so that there was enough space to put a small sheet of paper and write on it. He told me about three questions regarding smoothing $\Lambda$-sequences. He tried to answer them with somebody else earlier, but unsuccessfully, and recently he needed answers for his other math work. Dan showed me the problems and we discussed them for some time. The questions looked nice, but we found no answers. However, when I entered math department's office the next morning I was able to exclaim excitedly: "I found the answers overnight!" Dan smiled widely and replied: "So did I". Indeed, we both had somewhat distinct solutions, but all of them were correct, and we published a joint note on the problem later — my only joint paper with Dan.

In the early spring of 1994 I had my Ph.D. thesis ready, and for a number of weeks Dan invited me to his house regularly. He wanted to learn the details of my math and we went over the thesis very carefully. Some proofs were extremely technical, but Dan insisted on checking every detail, and we worked hard for hours. We found numerous mistakes and although none of them were serious, correcting them according to Dan's advice made my thesis readable. I do not remember the details of proofreading, but I do remember that one day Dan offered me a glass of whiskey. I was not a fan of the strange stuff and I explained to him that my experience with whiskey convinced me that the beverage was not invented for my taste. Dan smiled, took a bottle of Laphroaig and a glass of water, and poured two or three drops of the Islay single malt into the water. The smoky smell of Laphroaig filled the kitchen where we were working. I drank some of the whiskey tasting water and the distinct taste of the excellent single malt impressed me. (Every so often my second wife tells me that I'm too impressed even today!)

Dan kept helping me every time I needed help. He even sent me money when after my return to Poland and the sudden death of my first wife. I was in desperate financial condition. I haven't paid back the debt yet. I will, though, one day, not to Dan, but to somebody else who will need help. I learned that from Dan.

In March 2007 I visited Dan and Mudite in their sunny house in Florida. They invited us (me and my second wife) for a long stay at their place prior to the conference in honor of him at FAU. Some days Dan and I worked long hours on math, other days he cooked splendid dinners, baked superb bread or showed us what his favorite ice cream tastes like. We talked about all sorts of things, and I knew that I wanted to be like him when I would be his present age. The days spent with him and Mudite gave me strong inspiration for improving my lifestyle. I'm a bit happier every day going in their tracks but at my own pace.

The days gave me a deeper understanding of my relationship with Dan. I had known from the first weeks at SU that he was my mentor. Now I know that he is also the missing part of my father.

Franciszek Prus-Wisniowski

As my dissertation advisor, Dan Waterman guided me to the completion of my doctorate. He fulfilled this role excellently by making suggestions when I needed them, encouraging me when I was frustrated, and allowing me to pursue my own interests when I desired. He was also very helpful during the research process and would meet with me often. In fact, as I now advise graduate students on theses at the master's level, I try to follow the model he set.

However, Dan is not only a superior advisor, he is also a caring, friendly individual who concerns himself with the welfare of others. Since the completion of my degree, I continued to work with Dan, but have also become more friendly with him. In fact, each year when I am in the Boynton Beach area visiting family, I always make sure to stop and visit Dan. We will typically share a meal and some stories.

Paul Schembari

# ON CONCENTRATING IDEMPOTENTS, A SURVEY

J. MARSHAL ASH

*Mathematics Department, DePaul University*
*Chicago, IL 60614*
*E-mail: mash@math.depaul.edu*
*http://www.depaul.edu/~mash*

A sum of exponentials of the form $f(x) = \exp(2\pi i N_1 x) + \exp(2\pi i N_2 x) + \cdots + \exp(2\pi i N_m x)$, where the $N_k$ are distinct integers is called an *idempotent* trigonometric polynomial or, simply, an *idempotent* . It is known that for every $p > 1$, and every set $S$ of the torus $\mathbb{T} = \mathbb{R}/\mathbb{Z}$ with $|S| > 0$, there are idempotents concentrated on $S$ in the $L^p$ sense. We sketch how this concentration phenomenon originated as a reformulation of a functional analysis problem, and, in turn, studying concentration led to some interesting questions about $L^p$ norms of Dirichlet kernels associated with multiple trigonometric series. Some counterexamples involving linear operators not of convolution type are given.

*Keywords*: Concentration of idempotents, projections, weak type (2,2), restricted type (2,2), Dirichlet kernels, $L^p$norms, trigonometric polynomials

In 1977 I was visiting Stanford on sabbatical from DePaul. It was a most productive year, both personally and professionally. One the first side, I met and married Alison who subsequently gave me the second and third of my three wonderful sons. On the mathematical side, one of the best things was discussions with Mischa Zafran concerning a question about linear operators on $L^2(\mathbb{T})$.

## 1. From Operators on $L^2(\mathbb{Z})$ to Concentration

### 1.1. *Definitions*

By $L^2(\mathbb{Z}) = \ell^2$ we mean sequences $C = \{\ldots, c_{-1}, c_0, c_1, \ldots\}$ of complex numbers such that $\sum |c_\nu|^2 < \infty$. We identify the sequence with its Fourier series $C(x) = \sum c_\nu e^{2\pi i\nu x}$. A characteristic function associated to the finite subset $S = \{n_1, n_2, \ldots, n_K\}$ of $\mathbb{Z}$ is a sequence $\{\ldots, s_{-1}, s_0, s_1, \ldots\}$ where $s_\nu = 1$ when $\nu \in S$ and $s_v = 0$ when $\nu \notin S$. The product of $C = \{c_n\}$ and $D = \{d_n\}$ is $CD = \{c_n d_n\}$, while the convolution is $C{*}D = \left\{\sum_{\nu=-\infty}^{\infty} c_{n-\nu} d_\nu\right\}$. A fixed sequence $K = \{k_n\}$ creates an operator on functions on $\mathbb{Z}$ according to the rule $T : C \to K{*}C$. We identify $TC$ with the function $K(x)C(x)$ where $K(x) = \sum_{n=-\infty}^{\infty} k_n e^{2\pi i n x}$. By

Plancherel's formula, we have

$$\|TC\|_2^2 = \sum_{n=-\infty}^{\infty} \left| \sum_{\nu=-\infty}^{\infty} k_{n-\nu} c_\nu \right|^2 = \int_0^1 |K(x) C(x)|^2 \, dx.$$

To avoid confusion we will call the above mentioned characteristic functions *idempotents* and reserve the term *characteristic function* for a function of the form $\chi_E(x)$ which is 1 when $x \in E$, $E$ some measurable subset of $\mathbb{T} = [0,1]$, and 0 when $x \in \mathbb{T} \backslash E$. For each finite subset $S$ of $\mathbb{Z}$, the trigonometric polynomial associated to the idempotent associated to $S$ is $\iota_S(x) = \sum_{n \in S} e^{2\pi i n x}$; $\iota_S(x)$ derives its name from the identity $\iota_S \iota_S = \iota_S$. Here and henceforth we abuse notation and write $\iota$ for both the sequence and the associated trigonometric polynomial.

A linear operator defined on simple functions is *s.* $(2,2)$ or of *strong type* $(2,2)$ or *bounded on* $L^2(\mathbb{Z})$ if there is a constant $M > 0$ so that

$$\|TC\|_2 \leq M \|C\|_2$$

where $\|C\|_2 = \sqrt{\sum |c_n|^2} = \sqrt{\int_0^1 |C(x)|^2 \, dx}$. It is *r.* $(2,2)$ or of *restricted type* $(2,2)$ if there is a constant $M > 0$ such that

$$\|T\iota\|_2 \leq M \|\iota\|_2$$

whenever $\iota = \iota_S$ is an idempotent. It is *w.* $(2,2)$ or of *weak type* $(2,2)$ if there is a constant $M > 0$ such that

$$\|TC\|_{2\infty}^* \leq M \|C\|_2$$

where

$$\|C\|_{2\infty}^* := \sqrt{\sup_{\alpha > 0} |\{n \in \mathbb{Z} : |C^*(n)| > \alpha\}| \, \alpha^2}$$

and $C^*$ denotes the non-increasing rearrangement of $C$. Finally it is *w.r.* $(2,2)$ or of *weak restricted type* $(2,2)$ if there is a constant $M > 0$ so that for all idempotents

$$\|T\iota\|_{2\infty}^* \leq M \|\iota\|_2 .$$

### 1.2. *Relating classes of operators*

We have four trivial implications: simply restricting the action of $T$ to a subset of functions cannot increase the associated constant so that:

(1) if $T$ is *s.* $(2,2)$, then $T$ is *r.* $(2,2)$

and also

(2) if $T$ is *w.* $(2,2)$, then $T$ is *w.r.* $(2,2)$.

By Tchebycheff's inequality, for each $\alpha > 0$, $|\{n : |C^*(n)| > \alpha\}| \, \alpha^2 \leq \sum |c_n|^2$, so that:

(3) if $T$ is *s.* $(2,2)$, then $T$ is *w.* $(2,2)$

and also

(4) if $T$ is $r.\,(2,2)$, then $T$ is $w.r.\,(2,2)$.

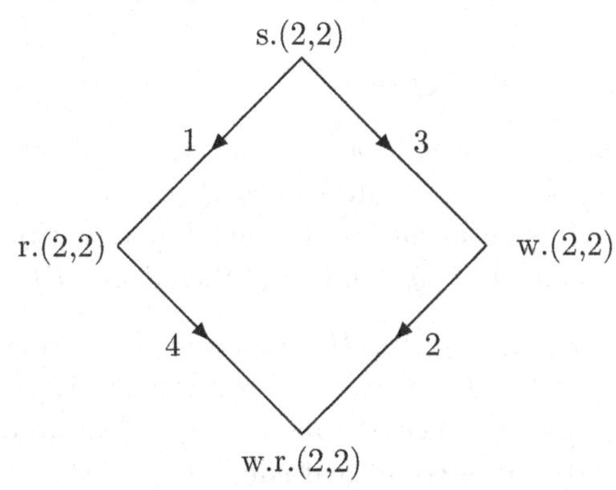


None of these implications are reversible in general. To see this, consider the following three linear operators initially defined on idempotents.

$$T_1(\{c_n\}) := \left\{ \left( \sum_{\nu=-\infty}^{\infty} \frac{c_\nu}{\sqrt{|\nu|+1}} \right) \frac{1}{|n|+1} \right\},$$

$$T_2(\{c_n\}) := \left\{ \left( \sum_{\nu=-\infty}^{\infty} \frac{c_\nu}{\sqrt{|\nu|+1}} \right) \frac{1}{\sqrt{|n|+1}} \right\}, \text{ and}$$

$$T_3(\{c_n\}) := \left\{ \left( \sum_{\nu=-\infty}^{\infty} \frac{c_\nu}{|\nu|+1} \right) \frac{1}{\sqrt{|n|+1}} \right\}.$$

The operators $T_1$ and $T_2$ cannot be defined on all of $\ell^2$; look at the sequence $\left\{ \frac{1}{\sqrt{|n|+1}\ln(|n|+2)} \right\} \in \ell^2$ to see this. Since $T_1 \in r.\,(2,2)$, but not $s.\,(2,2)$, implication (1) is irreversible. Since $T_2 \in w.r.\,(2,2)$, but not $w.\,(2,2)$, implication (2) is irreversible. Since $T_3 \in w.\,(2,2)$ and hence $w.r.\,(2,2)$, but not in $r.\,(2,2)$ and hence not in $s.\,(2,2)$; neither implication (3) nor implication (4) can be reversed. When the underlying group is the torus $\mathbb{T} = [0,1)$ with addition mod 1, there are similar counterexamples to all four implications. (See [SW] or [As1] for examples.)

Alexander Stokolos asked me if the fact that the example $T_1$ and $T_2$ are not defined on all of $L^2$ is crucial. For example, one might conjecture that a restricted $(2,2)$ linear operator defined on all of $L^2(\mathbb{T})$ is necessarily bounded. Paul Hagelstein and Brian Raines created the following counterexample for me.

**Theorem 1.1.** *There exists a linear operator $T : L^2(\mathbb{T}) \to \mathbb{R}$ such that $T\chi_E = 0$ for any measurable set $E$ and such that $T$ is unbounded on $L^2(\mathbb{T})$.*

**Proof.** Let $\mathcal{S}$ denote the set of simple functions on $\mathbb{T}$. Let $g_1 \in L^2/\mathcal{S}$ and $[g_1] = \{ag_1 + h : a \in \mathbb{C}, h \in \mathcal{S}\}$. Proceeding with transfinite induction, assume

$\{g_\gamma\}_{\gamma<\beta}$ and $\{[g_\gamma]\}_{\gamma<\beta}$ have been constructed. Let $g_\beta \in L^2/\bigcup_{\gamma<\beta}[g_\gamma]$ and set $[g_\beta] = \left\{ag_\beta + h : a \in \mathbb{C}, h \in \bigcup_{\gamma<\beta}[g_\gamma]\right\}$. Note $L^2 = \bigcup_{\gamma<2^{\omega_\circ}}[g_\gamma]$, where $2^{\omega_\circ}$ is the ordinality of the continuum. Let $\varphi$ be any bijection from the $\{g_\gamma : \gamma < 2^{\omega_\circ}\}$ to the real numbers. Define $Ts = 0$ for all $s \in \mathcal{S}$. If $f = ag_1 + h$ with $h \in \mathcal{S}$, define $Tf = a\|g_1\|_2\,\varphi(g_1)$; $T$ is linear on $[g_1]$. Extend $T$ inductively to $L^2$: if $T$ is defined and linear on $\bigcup_{\gamma<\beta}[g_\gamma]$ and $f = ag_\beta + h \in [g_\beta]$, then define $Tf = a\|g_\beta\|_2\,\varphi(g_\beta) + Th$. Now $T$ is defined and linear on $[g_\beta]$. By the principle of transfinite induction, $T$ is defined and linear on all of $L^2$. The operator norm of $T|\mathcal{S}$ is zero, so $T \in r.(2,2)$; but $T$ is unbounded on $L^2$, since $T$ stretches the $L^2$ norm of $g_\beta$ by $|\varphi(g_\beta)|$, and $\varphi(g_\beta)$ can be any (arbitrarily large) real number. □

In the late 1970s several people suspected that these implications were reversible for convolution operators. I focused on trying to reverse implication (1), in other words to prove:

For convolution operators defined on idempotents, $r.(2,2)$ implies $s.(2,2)$. (1)

Because of the following simple duality result, this would also reverse implication (3).

**Lemma 1.1.** *If a convolution operator $T : C \to K * C$ has the property that $r.(2,2) \Longrightarrow s.(2,2)$, then by duality it has the property $w.(2,2) \Longrightarrow s.(2,2)$.*

**Proof.** For finite sequences $C = \{c_n\}$ and $D = \{d_n\}$ we have

$$(C, TD) = \sum_{n=-\infty}^{\infty} c_n\overline{(TD)_n} \tag{2}$$

$$= \sum_{n=-\infty}^{\infty} c_n\overline{\left(\sum_{\nu=-\infty}^{\infty} k_{n-\nu}d_\nu\right)} \tag{3}$$

$$= \sum_{\nu=-\infty}^{\infty}\left(\sum_{n=-\infty}^{\infty} c_n k^*_{\nu-n}\right)\overline{d_\nu} = (T^*C, D) \tag{4}$$

where for every $n \in \mathbb{Z}$, $k^*_n = \overline{k_{-n}}$ and $T^*$ is the convolution operator corresponding to $\{k^*_n\}$. Let $K^*(x)$ be the Fourier series associated with $\{k^*_n\}$.

Let $T$ be $w.(2,2)$. This means that there is a constant $M$ so that

$$\|TC\|^*_{2\infty} \le M\|C\|_2\,. \tag{5}$$

We show that this implies that $T^*$ is $r.(2,2)$.

Let $\iota_S$ be the idempotent associated to a finite set $S$, $S \subset \mathbb{Z}$. By the definition of adjoint and Lorentz's generalization of Holder's inequality(see page 261 of [Hu]) we have

$$\begin{aligned}\|T^*\iota_S\|_2 &= \sup_{\|D\|_2 \le 1} |(K^* * \iota_S, D)| \\ &= \sup_{\|D\|_2 \le 1} |(\iota_S, K * D)| \\ &\le 2 \sup_{\|D\|_2 \le 1} \|\iota_S\|_{21}^* \|K * D\|_{2\infty}^* .\end{aligned}$$

Applying inequality (5) yields

$$\|T^*\iota_S\|_2 \le M \|\iota_S\|_{21}^* . \tag{6}$$

Letting $C^*$ denotes the non-increasing rearrangement of a sequence $C$, the following calculations hold

$$\|\iota_S\|_{21}^* := \frac{1}{2}\int \iota_S^*(x)\frac{dx}{\sqrt{x}} = \frac{1}{2}\int_0^{|S|}\frac{dx}{\sqrt{x}} = \sqrt{|S|} \text{ and} \tag{7}$$

$$\|\iota_S\|_2 = \sqrt{\int (\iota_S^*(x))^2\,dx} = \sqrt{\int_0^{|S|} dx} = \sqrt{|S|}.$$

From this and inequality (6) we have

$$\|T^*\iota_S\|_2 \le M \|\iota_S\|_2 , \tag{8}$$

so that $T^*$ is $r.\,(2,2)$. Our hypothesis now yields that $T^*$ is $s.\,(2,2)$. Finally if $T^*$ is $s.\,(2,2)$, so is $T$. □

### 1.3. *A surprising connection*

We say that $L^2$ *interval concentration* occurs if there is an absolute constant $a > 0$ such that for each interval $I \subset [0,1]$ there is an idempotent $\iota(x) = \iota_I(x) = \sum_{j=1}^{K} e^{2\pi n_j x} \in L^2(\mathbb{Z})$ so that

$$\frac{\int_I |\iota(x)|^2\,dx}{\int_0^1 |\iota(x)|^2\,dx} > a$$

and that $L^2$ *set concentration* occurs if there is an absolute constant $b > 0$ such that for each set of positive measure $E \subset [0,1]$ there is an idempotent $\iota(x) = \iota_E(x)$ so that

$$\frac{\int_E |\iota(x)|^2\,dx}{\int_0^1 |\iota(x)|^2\,dx} > b. \tag{9}$$

Note that if $b$ exists we may take $a$ to be $b$.

When I was studying implication (1) in the late 1970s, I was only able to find an equivalent formulation in terms of concentration. Here is that equivalence.

**Theorem 1.2.** *If $L^2$ concentration for sets holds, then r. $(2,2)$ implies s. $(2,2)$ when the underlying group is $\mathbb{Z}$.*

*If r. $(2,2)$ implies s. $(2,2)$ when the underlying group is $\mathbb{Z}$, then $L^2$ concentration for intervals holds.*

**Proof.** Assume that $L^2$ concentration holds for sets and that $T$ is r. $(2,2)$. Letting $T$ correspond to convolution with $\{k_n\}$ so that with $K(x) = \sum k_n e^{2\pi i n x}$ for a certain positive constant $A$ the inequality

$$\int_0^1 |K(x)\,\iota(x)|^2\,dx \le A \int_0^1 |\iota(x)|^2\,dx \tag{10}$$

holds for every idempotent $\iota$. We also know that there is a positive number $b$ so that inequality (9) holds. Our goal is to show that $K(x)$ is an essentially bounded function, since this is well known to be the necessary and sufficient condition for a multiplier operator to be bounded on $L^2$. Assume that the multiplier $K(x)$ exceeds $A/b$ on a set $E$ of positive measure. Find an idempotent $\iota$ so that

$$\int_E |\iota(x)|^2\,dx > b \int_0^1 |\iota(x)|^2\,dx$$

Applying first this and then inequality (10), we find

$$\begin{aligned}\int_0^1 |\iota(x)|^2\,dx &< \frac{1}{b}\int_E |\iota(x)|^2\,dx\\ &= \frac{1}{A}\frac{A}{b}\int_E |\iota(x)|^2\,dx\\ &\le \frac{1}{A}\int_E |K(x)\,\iota(x)|^2\,dx\\ &\le \frac{1}{A}A\int_0^1 |\iota(x)|^2\,dx,\end{aligned}$$

which is a contradiction.

The converse implication is Theorem 7 of [As1]: Assume the failure of $L^2$ concentration for intervals. Then there is a sequence of intervals $\{I_i\}_{i=1,2,\ldots}$ so that for every $i$ and every idempotent $\iota$,

$$\int_{I_i} |\iota(x)|^2\,dx \le 2^{-2i} \int_0^1 |\iota(x)|^2\,dx.$$

Then $K(x) = \sum i\chi_{I_i}(x)$ is unbounded so that the operator corresponding to the multiplier $K$ is not bounded on $L^2(\mathbb{Z})$. However, for any idempotent $\iota$, by

Minkowski's inequality we have

$$\begin{aligned}
\|K(x)\,\iota(x)\|_2 &\leq \sum\left(\int_0^1 |i\chi_{I_i}(x)\,\iota(x)|^2\,dx\right)^{1/2} \\
&= \sum i\left(\int_{I_i} |\iota(x)|^2\,dx\right)^{1/2} \\
&\leq \left(\sum i2^{-i}\right)\left(\int_0^1 |\iota(x)|^2\,dx\right)^{1/2} \\
&= 2\,\|\iota(x)\|_2
\end{aligned}$$

so that the operator *is* $r.\,(2,2)$. □

## 1.4. *Results for $L^2$ Concentration*

Just about the time of the formulation of the equivalence theorem, Michael Cowling proved that when the underlying group is $\mathbb{Z}$, $r.\,(2,2)$ implies $s.\,(2,2)$. [Co] Actually, he proved much, much more than this. He proved that if the underlying group is any amenable group, than $w.r.\,(2,2)$ implies $s.\,(2,2)$. An amenable group is a topological group G carrying a kind of averaging operation, that is invariant under translations by group elements. In the case where G is not an abelian group, that means translation on a fixed side (left- or right-translation). For our purposes, it is enough to know that $\mathbb{Z}$ is an amenable group. So by the equivalence theorem above, it followed that $L^2$ concentration for intervals was true!

But this result is exceedingly non constructive. Define the absolute constant $C_2$ as the largest real number such that for every set $E \subset \mathbb{T}$ with $|E| > 0$, and every $\epsilon > 0$, there is an idempotent $\iota = \iota_{E,\epsilon}$ satisfying the inequality

$$\frac{\sqrt{\int_E |\iota(x)|^2 dx}}{\sqrt{\int_{\mathbb{T}} |\iota(x)|^2 dx}} \geq C_2 - \epsilon. \tag{11}$$

Thus $C_2$ is the amount of $L^2$ norm that can be concentrated on any set, no matter how small, nor no matter how inconveniently situated. Obviously, $C_2 \leq 1$. So far we know that $C_2 > 0$, but our information is neither quantitative nor constructive, we have no idea of its size, no lower bound, nor any effective procedure for finding one.

## 1.5. *Quantitative results for $L^2$ concentration*

(1) The referee of paper [As1] pointed out that $C_2$ must be at least $1/8 = .125$. This follows from Cowling's Theorem. The assumption of Cowling's Theorem, that $T$ is a convolution operator of type $w.r.\,(2,2)$ means that there is a constant $A > 0$ such that for every idempotent $\iota$, we have

$$\|T\iota\|_{2\infty}^* \leq A\,\|\iota\|_2 = A\,\|\iota\|_{21}^*. \tag{12}$$

There are two steps to the proof that $T$ is of strong type; first we show that inequality (12) can be extended to hold for simple functions and hence for all functions in $L^{21}$ - this step results in the operator norm being stretched by at most 8. In other words, producing the inequality

$$\|TC\|_{2\infty}^{*} \leq 8A \|C\|_{21}^{*}$$

for every sequence $C \in L^{21}$. (This step is shown in detail on page 682 of [As1].) The second step gets from this to the final result

$$\|TC\|_{2} \leq 8A \|C\|_{2} \tag{13}$$

for every $C \in L^{2}$ with no further increase in operator norm.

Assume that the $L^2$ concentration constant $C_2$ satisfies $C_2 < 1/8$. This means that there is a set $E \subset \mathbb{T}$ of positive measure so that for every idempotent $\iota$,

$$\|\chi_E(x)\,\iota(x)\|_2 \leq C_2 \|\iota(x)\|_2 . \tag{14}$$

Let $T$ be the linear operator associated with the multiplier function $\chi_E(x)$. Then by Tchebycheff's inequality, for any idempotent $\iota$

$$\|T\iota\|_{2\infty}^{*} \leq \|T\iota\|_2 = \|\chi_E(x)\,\iota(x)\|_2 . \tag{15}$$

Concatenating inequalities (14) and (15) gives

$$\|T\iota\|_{2\infty}^{*} \leq C_2 \|\iota(x)\|_2 .$$

Thus by the remark above that the strong $(2,2)$ constant is at most 8 times the weak restricted $(2,2)$ constant we have that for every sequence $C$,

$$\|TC\|_2 \leq 8C_2 \|C\|_2 .$$

But $8C_2 < 1$, which contradicts the fact that the strong $(2,2)$ norm of the operator $T$ must be 1, since 1 is the essential supremum of the multiplier function $\chi_E$.

Define a constant $C_2^*$ which is for intervals what $C_2$ is for sets. In other words, the absolute constant $C_2^*$ is the largest real number such that for every interval $J \subset \mathbb{T}$ with $|J| > 0$, and every $\epsilon > 0$, there is an idempotent $\iota = \iota_{J,\epsilon}$ satisfying the inequality

$$\frac{\sqrt{\int_J |\iota(x)|^2 dx}}{\sqrt{\int_{\mathbb{T}} |\iota(x)|^2 dx}} \geq C_2^* - \epsilon. \tag{16}$$

Of course $C_2^* \geq C_2$.

(2) S. Pichorides [Pi] obtained $C_2^* \geq .14$.

(3) H. L. Montgomery [Mo], and

(4) J.-P. Kahane [Ka2] obtained several better lower bounds.

(The ideas of H. L. Montgomery were "deterministic" while those of J.-P. Kahane used probabilistic methods from [Ka1].)

### 1.6. *The best possible $L^2$ concentration constant*

Finally, in [AJS], together with Roger Jones and Bahman Saffari, I achieved this lower bound for $C_2$ :

$$\gamma_2 := \max_{x>0} \frac{\sin x}{\sqrt{\pi x}} = 0.4802\ldots; \tag{17}$$

which, in [DPQ1], was proved to be best possible, thus $C_2 = \gamma_2$. (See [DPQ2] for a more detailed exposition of the contents of [DPQ1].)

## 2. A Paper 20 Years in the Making

In 1982, I began to consider the $L^p$ concentration question for values of $p$ other than 2. This question represents a move away from the functional analysis issues naturally connected to the $L^2$ concentration question in Theorem 1.2 for two reasons. First, only for $p = 2$ do we have the very simple characterization of a convolution operator being bounded if and only if the corresponding multiplier function is in $L^\infty$. Second, Misha Zafran has shown that even when the underlying group is $\mathbb{T}$ so that things are as simple as possible, there are convolution operators of type $w.\,(p,p)$ but not $s.\,(p,p)$ when $1 < p < 2$ and thus by duality there are also convolution operators of type $r.\,(p,p)$ but not $s.\,(p,p)$ when $2 < p < \infty$. [Za] Consequently, $w.r.\,(p,p)$ implies $s.\,(p,p)$ only when $p = 2$.

Misha Zafran was a talented mathematician. When I was spending 1977 at Stanford, Misha selflessly and patiently shared many ideas. The subject discussed here owes much of its origin to Misha. I profoundly regret his untimely death.

### 2.1. *The early years*

In the fall of 1982, Roger Jones and I submitted a grant proposal to the National Science Foundation centered around the question of whether $L^p$ concentration was valid for any $p < 2$. One proposal reviewer wrote "...This [$L^p$ concentration] is a very specific problem. They [Ash and Jones] mention several ways it has been done for $p > 2$. Doing it for $p < 2$ doesn't seem very difficult either; a product of Dirichlet kernels is likely to work." Needless to say, the proposal was not funded. Intrigued by the above comment, I communicated a desire to the NSF analysis director John Ryff that the referee divest anonymity and collaborate on the question. A few months later, I received a letter from Dan Rider stating "...Enclosed is a sketch of what I think works for your problem for $1 < p \leq 2$. It turned out to be messier than I had anticipated..."

Here, in very heuristic terms, is Rider's idea. We fix $p > 1$ and explain how to find an idempotent which has a goodly percentage of its $L^p$ mass near $k/q$, where $q$ is a large prime and $1 \leq k \leq q - 1$. First let $k = 1$. We concentrate $L^p$ mass on the interval $\left[\frac{1}{q}-\frac{1}{q^2},\frac{1}{q}+\frac{1}{q^2}\right]$ by considering the idempotent

$$I(x)=D_{q^2}(qx)\,D_{\frac{q-1}{2}}(x),$$

where $D_n(x)=\sum_{\nu=0}^{n-1}e^{2\pi i\nu x}$. The first factor is concentrated near $x=0$ and has period $1/q$ so we think of it as being roughly

$$c\sum_{j=0}^{q-1}\chi_{\left[\frac{j}{q}-\frac{1}{q^2},\frac{j}{q}+\frac{1}{q^2}\right]}(x),$$

in other words as a series of $q$ equal pulses. The second factor we think of as being even and having decay like $1/x$ on $[1/q,1/2]$. This gives $L^p$ concentration since

$$\frac{\int_{\frac{1}{q}+\frac{1}{q^2}}^{\frac{1}{q}-\frac{1}{q^2}}|I(x)|^p\,dx}{\int_0^1|I(x)|^p\,dx}\approx\frac{\sum_{j=1}^{1}\frac{1}{j^p}}{\sum_{j=1}^{(q-1)/2}\frac{1}{j^p}}\geq\frac{1}{\sum_{j=1}^{\infty}\frac{1}{j^p}}.$$

If $k>1$, since the set $\{1,2,\ldots,q-1\}$ is a group under multiplication modulo $q$, we may find $a$ so that $ka\equiv 1 \bmod q$ and use

$$D_{q^2}(qx)\,D_{\frac{q-1}{2}}(ax)$$

as our concentrated idempotent. Notice that $a\left(\frac{k}{q}+\delta\right)\equiv\frac{1}{q}+a\delta \bmod q$ so that this idempotent behaves at $k/q$ in a way that is similar to how $I(x)$ behaved at $1/q$.

## 2.2. *On the virtues of procrastination*

Our early work on the $L^p$ concentration problem is summarized in [AAJRS1]. This paper lists the results of what we were able to prove developing what was mentioned above. We had established $L^p$ concentration for intervals for $1<p<\infty$ and $L^p$ concentration for sets for $2\leq p<\infty$. A good thing was that the method was both quantitative and constructive. But there was a 17 year gap between the $L^2$ result in [AJS] and this and there would be another 7 year gap between this and our next paper [AAJRS2]. Two reasons for this time gap were our desire to find out whether $L^p$ concentration for sets held when $1<p<2$ and whether there was any $L^1$concentration. We had a little negative evidence for the latter and more interesting (to us) of these questions, namely that if one considers the "enemy" interval $\left[\frac{1}{q}-\frac{1}{q^2},\frac{1}{q}+\frac{1}{q^2}\right]$, the fraction of the $L^1$ mass of $I(x)$ concentrated here is roughly

$$\frac{1}{\sum_{j=1}^{(q-1)/2}\frac{1}{j}}\approx\frac{1}{\ln q}.$$

So if $I(x)$ were about as concentrated as an idempotent can be, then $C_1$ would be less than $\frac{1}{\ln q}$ for arbitrarily large $q$ and $L^1$ concentration would be false.

We finally gave up on resolving these two questions and sent the paper to Annal. Inst. Fourier where an excellent referee showed us how to solve the former question, that is how to establish concentration for sets when $1 < p < 2$. What was surprising to me was that doing this involved using the following two dimensional result.

**Lemma 2.1 (Triangle Lemma).** *Let $p > 1$, $0 < \theta < 1$, and*

$$K_{\theta,N} = \left\{(x,y) \in \mathbb{Z}^2 : x + \theta^{-1}y \leq N, x \geq 0, y \geq 0\right\}.$$

*Then for arbitrary $N \geq 4$,*

$$\int_0^1 \int_0^1 \left| \sum_{(m,n)\in K_{\theta,N}} e^{2\pi i(mx+ny)} \right|^p dxdy \leq C_p N^{2p-2} \tag{18}$$

*uniformly with respect to $\theta$ and $N$.*

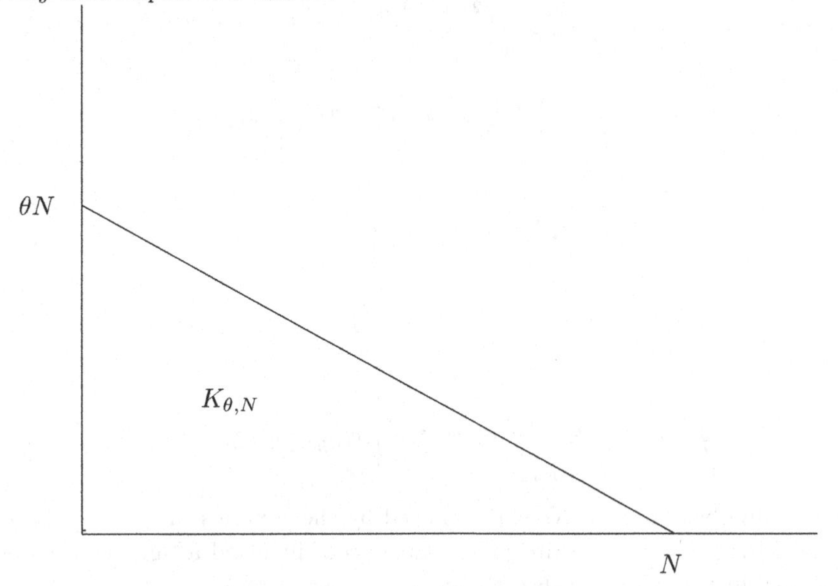


But this was still not the end of the line for paper [AAJRS2]. We resubmitted the paper with the Triangle Lemma linked in, but the referee then pointed out that the proof of the Triangle Lemma was just fine, but that the linkage was defective. We now needed a number theory fact that none of us knew, namely:

**Lemma 2.2.** *Almost every point $\xi$ has the property that there are infinitely many primes $q$ and integers $k$ for which*

$$\left|\xi - \frac{k}{q}\right| < \frac{1}{q^2}.$$

But paper [AAJRS2] had taken so long to evolve that the internet was now equipped with strong enough search engines to allow a one day wrap up. Searching for "Approximation to Irrational Number by Rational Numbers with Prime Denominators" led me directly to Chao-Hua Jia in China. His immediate response to my email in turn led me to Glyn Harman in England. Harman immediately emailed me that Lemma 2.2 was on page 27 of his book! [Ha] One last point: we were able to guess the identity of the referee, from the combination of his previously displayed expertise in moving from intervals to sets of positive measure, his involvement with two dimensional estimates like the Triangle Lemma, and his extreme generosity. Since Fedja Nazarov would not let us make him a coauthor, I thank him here.

## 3. The Future

### 3.1. *A segue*

The Triangle Lemma, Lemma 2.1 was a natural conjecture since

$$\begin{aligned}
&\int_0^1\int_0^1\left|\sum_{|m|,|n|\le N} e^{2\pi i(mx+ny)}\right|^p dxdy \\
&=\int_0^1\int_0^1\left|\sum_{|m|\le N} e^{2\pi imx}\sum_{|n|\le N} e^{2\pi iny}\right|^p dxdy \\
&=\left\{\int_0^1\left|\sum_{|m|\le N} e^{2\pi imx}\right|^p dx\right\}^2 \\
&=\left\{O\left(N^{p-1}\right)\right\}^2 = O\left(N^{2p-2}\right).
\end{aligned}$$

Furthermore,

$$\int_0^1\int_0^1\left|\sum_{(m,n)\in K_{\theta,N}} e^{2\pi(mx+ny)}\right| dxdy \le C_p \ln^2 N,$$

(which is again obvious when $K_{\theta,N}$ is replaced by the square $\{|m|,|n|\le N\}$) *had* been proved by Yudin and Yudin. [YY] Their proof involved a number theoretic argument and luckily extended directly to $p>1$. (See [As2].)

Thinking further about the Triangle Lemma led me to think about generalizations. First of all, every convex polygon can be decomposed into a finite number of triangles, so it is almost immediate that an estimate like (18) holds for convex polygons in general. The next thought was to try to push the result up to three dimensions. Unfortunately, the number theory required to extend the method to a higher dimension appeared formidable.

The one dimensional integral

$$\int_0^1\left|\sum_{|k|\le N} e^{2\pi ikx}\right| dx \backsimeq \frac{4}{\pi^2}\ln N$$

is called the "Lebesgue constant ". The $L^p$ Lebesgue constant is

$$\int_0^1\left|\sum_{|k|\le N} e^{2\pi ikx}\right|^p dx \backsimeq \delta_p N^{p-1}.$$

Since $\{k \in \mathbb{Z} : |k| \leq N\} = N[-1,1]$ is the set of lattice points in the dilate of the one dimensional convex polygon $[-1,1]$ by $N$, it is natural to let

$$L_N(D) := \int_{\mathbb{T}^d} \left| \sum_{k \in ND \cap \mathbb{Z}^d} e^{2\pi i k \cdot x} \right|^p dx$$

be a $d$-dimensional $L^p$ Lebesgue constant for any bounded set $D \subset \mathbb{R}^d$. This raised the following two questions. (1) If $p > 1$ and $D$ is a $d$-dimensional polyhedron, do there exist constants $c = c(p, D)$ and $C = C(p, D)$ so that

$$cN^{d(p-1)} \leq L_N(D) \leq CN^{d(p-1)}?$$

(2) What can be said for more general bounded $d$-dimensional sets with nonempty interior?

The first question became much more accessible when Elijah Liflyand told me that Belinsky had proved the $p = 1$ analogue of this with a methodology that avoided the delicate number theory completely [BT]. With respect to the second question, here is a conjecture.

**Conjecture 3.1.** *If $D$ is a bounded $d$-dimensional set with nonempty interior then*

$$cN^{d(p-1)} \leq L_N(D) \leq \begin{cases} CN^{d(p-1)} & \text{if } p > \frac{2d}{d+1} \\ CN^{\frac{d-1}{2}p} \ln^{\frac{d+1}{2d}p} N & \text{if } p = \frac{2d}{d+1} \\ CN^{\frac{d-1}{2}p} & \text{if } 1 < p < \frac{2d}{d+1} \end{cases}.$$

Laura De Carli and I have affirmed the first question and have some partial progress that points toward the conjecture in a recently submitted paper. [AD]

### 3.2. *The $L^1$ concentration question*

Aline Bonami and Szilárd Gy. Révész appear to have disproved the main conjecture of [AAJRS2]. Evidentially $L^p$ concentration *does* occur for $p = 1$ and even for some $p < 1$ as well. These results should appear soon. To keep abreast of developments, look at Révész's website, http://www.renyi.hu/~revesz/preprints.html.

### 3.3. *A conjecture about operators*

A long standing question about convolution operators is this.

**Conjecture 3.2.** *Let $p > 2$. If a convolution operator is w. $(p,p)$, then it must also be s. $(p,p)$.*

I have no feelings either way concerning the truth of this conjecture. Misha Zafran warned me of difficulty in 1977, so I am not too surprised to see that it is still unsolved.

## References


[AAJRS1] B. Anderson, J. M. Ash, R. L. Jones, D. G. Rider, and B. Saffari, *$L^p$ norm local estimates for exponential sums,* *C. R. Acad. Sci. Paris Ser. I Math.* **330** (2000), 765–769.

[AAJRS2] B. Anderson, J. M. Ash, R. L. Jones, D. G. Rider and B. Saffari, *Exponential sums with coefficients 0 or 1 and concentrated $L^p$ norms, Annal. Inst. Fourier,* **57**(2007), 1377-1404.

[AJS] J. M. Ash, R. L. Jones and B. Saffari, *Inégalités sur des sommes d'exponentielles, C. R. Acad. Sci. Paris Ser. I Math.* **296** (1983), 899–902.

[As1] J. M. Ash, *Weak restricted and very restricted operators on $L^p$*, *Trans. Amer. Math. Soc.* **281** (1984), 675–689.

[As2] J. M. Ash ,*Triangular Dirichlet kernels and growth of $L^p$ Lebesgue constants, preprint,*`http://condor.depaul.edu/\symbol{126}mash/YudinLp.pdf`, (2004).

[AD] J. M. Ash and L. De Carli, *Growth of $L^p$ Lebesgue constants for convex polyhedra and other regions, Trans. Amer. Math. Soc.,* to appear. Also see `http://condor.depaul.edu/\symbol{126}mash/projectir2'.pdf`, (2007).

[BT] E. S. Belinsky and R. M. Trigub, *Fourier Analysis and Approximation of Functions,* (Kluwer Academic Publishers, Dordrecht, 2004).

[Co] M. Cowling , *Some applications of Grothendieck's theory of topological tensor products in harmonic analysis, Math. Ann.* **232** (1978), 273–285.

[DPQ1] M. DéChamps-Gondim, F. Piquard-Lust and H. Queffelec, *Estimations locales de sommes d'exponentielles, C. R. Acad. Sci. Paris Ser. I Math.* **297** (1983), 153–157.

[DPQ2] ——, *Estimations locales de sommes d'exponentielles , Université de Paris-Sud, Equipe de recherche associée au curs (296) analyse harmonique Mathématique (Bât. 425), Exposé,* **1** (1982–1983), 1–16.

[Ha] G. Harman, *Metric Number Theory,* (Clarendon Press, Oxford, 1998), p. 27, Cor. 2.

[Hu] R. A. Hunt, *On $L(p,q)$ spaces, Enseignement Math.* (2) **12** (1966), 249–276.

[Ka1] J.-P. Kahane, *Some random series of functions*, (2nd edn., Cambridge Studies in Advanced Math., 5, Cambridge Univ. Press, Cambridge–New York, 1985).

[Ka2] ——, *personal communication, November* (1982).

[Mo] H. L. Montgomery ,*personal communications, November 1980 and October* (1982).

[Pi] S. K. Pichorides,*personal communication, April* (1980).

[SW] E. M. Stein and G. Weiss, *An extension of a theorem of Marcinkiewicz and some of its applications, Indiana Univ. Math. J.* **8** (1959), 263-284.

[YY] A. A. Yudin and V. A. Yudin, *Polygonal Dirichlet kernels and growth of Lebesgue constants,* (Russian) *Mat. Zametki* **37** (1985) 220–236. {English translation: *Math. Notes* **37** (1985), 124–135.}

[Za] M. Zafran, *Multiplier transformations of weak type* , *Ann. of Math.* **101** (1975), 34–44.

# VARIANTS OF A SELECTION PRINCIPLE FOR SEQUENCES OF REGULATED AND NON-REGULATED FUNCTIONS

VYACHESLAV V. CHISTYAKOV

*Department of Applied Mathematics, State University Higher School of Economics, Bol'shaya Pechërskaya Street 25/12, Nizhny Novgorod 603155, Russia, czeslaw@mail.ru*

CATERINA MANISCALCO

*Dipartimento di Matematica ed Applicazioni, Università degli Studi di Palermo, Via Archirafi 34, 90123 Palermo, Italy, maniscal@math.unipa.it*

YULIYA V. TRETYACHENKO

*Department of Mathematics and Mechanics, University of Nizhny Novgorod, Gagarin Avenue 23, Nizhny Novgorod 603950, Russia, tretyachenko_y_v@mail.ru*

## Dedicated to Professor Daniel Waterman on the occasion of his 80th Birthday

Let $T$ be a nonempty subset of $\mathbb{R}$, $X$ a metric space with metric $d$ and $X^T$ the set of all functions mapping $T$ into $X$. Given $\varepsilon > 0$ and $f \in X^T$, we denote by $N(\varepsilon, f, T)$ the least upper bound of those $n \in \mathbb{N}$, for which there exist numbers $s_1, \ldots, s_n, t_1, \ldots, t_n$ from $T$ such that $s_1 < t_1 \leq s_2 < t_2 \leq \cdots \leq s_n < t_n$ and $d(f(s_i), f(t_i)) > \varepsilon$ for all $i = 1, \ldots, n$ ($N(\varepsilon, f, T) = 0$ if there are no such $n$'s). The following pointwise selection principle is proved: *If a sequence of functions $\{f_j\}_{j=1}^{\infty} \subset X^T$ is such that the closure in $X$ of the sequence $\{f_j(t)\}_{j=1}^{\infty}$ is compact for each $t \in T$ and $\limsup_{j\to\infty} N(\varepsilon, f_j, T) < \infty$ for all $\varepsilon > 0$, then $\{f_j\}_{j=1}^{\infty}$ contains a subsequence, converging pointwise on $T$ to a function $f \in X^T$, such that $N(\varepsilon, f, T) < \infty$ for all $\varepsilon > 0$.* We establish several variants of this result for functions with values in a metric semigroup and reflexive separable Banach space as well as for the weak pointwise and almost everywhere convergence of extracted subsequences, and comment on the necessity of conditions in the selection principles. We show that many Helly-type pointwise selection principles are consequences of our results, which can be applied to sequences of non-regulated functions, and compare them with recent results by Chistyakov [J. Math. Anal. Appl. 310 (2005) 609–625] and Chistyakov and Maniscalco [J. Math. Anal. Appl. 341 (2008) 613–625].

*Keywords*: Pointwise convergence, selection principle, regulated function, generalized variation, metric space, metric semigroup, Banach space, double sequence, weak convergence, almost everywhere convergence

## 1. Regulated Functions and Selection Principles

The aim of this section is to exhibit a certain relationship between characterizations of regulated functions and pointwise selection principles, having to do with the existence of pointwise convergent subsequences of a given sequence of functions, and to expose the main goal of the paper.

Given a closed interval $I = [a, b]$ on the real line $\mathbb{R}$ with $a < b$, denote by $\mathrm{Reg}(I)$ the set of all *regulated* functions $f : I \to \mathbb{R}$, i. e., those functions for which the left limit $f(t-) \in \mathbb{R}$ exists at each point $a < t \le b$ and the right limit $f(t+) \in \mathbb{R}$ exists at each point $a \le t < b$. The set $\mathrm{Reg}(I)$ is of importance, e. g., in the theory of everywhere convergence of Fourier series and in the theory of stochastic processes, and it is well known that each $f \in \mathrm{Reg}(I)$ is bounded, has at most a countable set of points of discontinuity, and is the uniform limit of a sequence of step functions.

The historically first selection principle was given by Helly [27] in the class $\mathrm{Mon}(I) \subset \mathrm{Reg}(I)$ of all monotone functions on $I$: *a uniformly bounded sequence* $\{f_j\} \equiv \{f_j\}_{j=1}^{\infty} \subset \mathrm{Mon}(I)$ *contains a pointwise convergent subsequence whose pointwise limit belongs to* $\mathrm{Mon}(I)$.

There are several generalizations of the Helly selection principle. In order to recall some of them that are relevant to our exposition, we introduce some notation.

Given a positive integer $n \in \mathbb{N}$ and $\varnothing \ne T \subset \mathbb{R}$, we denote by $\{I_i\}_1^n \prec T$ a collection of $n$ two-point sets of the form $I_i = \{s_i, t_i\} \subset T$, $i = 1, \dots, n$, for which $s_1 < t_1 \le s_2 < t_2 \le \dots \le s_{n-1} < t_{n-1} \le s_n < t_n$. In a different interpretation $\{I_i\}_1^n \prec T$ may be understood as a collection of $n$ ordered non-overlapping intervals $I_i = [s_i, t_i] \subset \mathbb{R}$ whose endpoints $s_i$ and $t_i$ lie in $T$, $i = 1, \dots, n$. A notation of the form $I \prec T$ will be used to mean that $I = \{a, b\}$ or $I = [a, b]$ with $a, b \in T$ and $a < b$.

If $f : I = [a, b] \to \mathbb{R}$ and $I_i = [s_i, t_i] \in \{I_i\}_1^n \prec I$, we set $|f(I_i)| = |f(s_i) - f(t_i)|$.

Given a nondecreasing continuous function $\varphi : \mathbb{R}^+ \equiv [0, \infty) \to \mathbb{R}^+$ vanishing at zero only and such that $\varphi(u) \to \infty$ as $u \to \infty$, called a *$\varphi$-function*, a function $f : I \to \mathbb{R}$ is said to be *of bounded $\varphi$-variation on $I$* in the sense of Wiener [38] and Young [39], in which case we write $f \in \mathrm{BV}_\varphi(I)$, provided the *$\varphi$-variation* of $f$ defined by

$$V_\varphi(f) = \sup\Big\{\sum_{i=1}^{n} \varphi(|f(I_i)|) : n \in \mathbb{N} \text{ and } \{I_i\}_1^n \prec I\Big\} \text{ is finite.}$$

In the particular case when $\varphi(u) = u$ the value $V_\varphi(f)$ is the usual Jordan variation of $f$, which will be written as $V(f)$; in this case we will also write $\mathrm{BV}(I)$ in place of $\mathrm{BV}_\varphi(I)$. It is well known (Young [39], Musielak and Orlicz [29]) that $\mathrm{BV}_\varphi(I) \subset \mathrm{Reg}(I)$, and if $\varphi$ is convex and $\varphi(u)/u \to 0$ as $u \to +0$, then $\mathrm{BV}(I)$ is a proper subset of $\mathrm{BV}_\varphi(I)$. On the other hand, the following characterization of the set $\mathrm{Reg}(I)$ was given by Goffman, Moran and Waterman [25]: if $f \in \mathrm{Reg}(I)$ and $\min\{f(t-), f(t+)\} \le f(t) \le \max\{f(t-), f(t+)\}$ at each point $t \in I$ of discontinuity of $f$, then $f \in \mathrm{BV}_\varphi(I)$ for some convex $\varphi$-function $\varphi$ such that $\varphi(u)/u \to 0$ as $u \to +0$.

The following generalizations of the Helly selection principle are known due to Helly [27] (if $* = \sqcup$) and Musielak and Orlicz [29] (if $* = \varphi$ is a $\varphi$-function):

**Theorem A.** *A pointwise bounded sequence of functions $\{f_j\}$ mapping $I$ into $\mathbb{R}$ and satisfying the condition*

$$\sup_{j\in\mathbb{N}} V_*(f_j) < \infty \tag{1}$$

*contains a pointwise convergent subsequence whose pointwise limit $f \in \mathrm{BV}_*(I)$.*

A sequence $\Lambda = \{\lambda_i\}_{i=1}^{\infty}$ of positive numbers is said to be a *Waterman sequence* [36] if it is nondecreasing, unbounded and the series $\sum_{i=1}^{\infty} 1/\lambda_i$ diverges. A function $f : I \to \mathbb{R}$ is said to be *of $\Lambda$-bounded variation on $I$* in the sense of Waterman [36, 37] (see also Goffman, Nishiura and Waterman [26, Section 11.3]), in which case we write $f \in \Lambda\mathrm{BV}(I)$ or $f \in \mathrm{BV}_\Lambda(I)$, provided the $\Lambda$-*variation* of $f$ given by

$$V_\Lambda(f) = \sup\Big\{\sum_{i=1}^{n} \frac{|f(I_i)|}{\lambda_{\sigma(i)}} : n \in \mathbb{N},\ \{I_i\}_1^n \prec I \text{ and } \sigma \in S_n\Big\} \text{ is finite,}$$

where $S_n$ is the set of all permutations $\sigma : \{1, \ldots, n\} \to \{1, \ldots, n\}$. It is well known (Waterman [36]) that $\mathrm{BV}_\Lambda(I) \subset \mathrm{Reg}(I)$ and $\mathrm{BV}(I)$ is a proper subset of $\mathrm{BV}_\Lambda(I)$. On the other hand, Perlman [30] showed that $\bigcap_\Lambda \mathrm{BV}_\Lambda(I) = \mathrm{BV}(I)$ and presented the following characterization of the set of all regulated functions: $\mathrm{Reg}(I) = \bigcup_\Lambda \mathrm{BV}_\Lambda(I)$, the intersection and the union being taken over all Waterman sequences $\Lambda$. In [37] Waterman proved a generalization of the Helly selection principle in the class $\mathrm{BV}_\Lambda(I)$ which is exactly Theorem A with $* = \Lambda$.

Since $\mathrm{Mon}(I) \subset \mathrm{Reg}(I)$ and $\mathrm{BV}_*(I) \subset \mathrm{Reg}(I)$ for $* = \sqcup, \varphi$ or $\Lambda$ as above, the Helly selection principle and Theorem A are selection principles in the class of regulated functions.

Yet there are characterizations of regulated functions not having to do with the notions of bounded variations given above. One of them is due to Chanturiya [3, 4] (see also Goffman, Nishiura and Waterman [26, Section 11.3.7]), who introduced a notion of the *modulus of variation* $\{\nu(n, f, I)\}_{n=1}^{\infty}$ of $f : I \to \mathbb{R}$ by

$$\nu(n, f, I) = \sup\Big\{\sum_{i=1}^{n} |f(I_i)| : \{I_i\}_1^n \prec I\Big\}, \qquad n \in \mathbb{N}, \tag{2}$$

and showed that $\mathrm{Reg}(I) = \{f : I \to \mathbb{R} \mid \nu(n, f, I) = o(n)\}$, where the condition $\mu(n) = o(n)$ means that $\mu(n)/n \to 0$ as $n \to \infty$. Making use of this notion, in [11, 12] Chistyakov replaced condition (1) by a very weak condition

$$\limsup_{j\to\infty} \nu(n, f_j, I) = o(n) \tag{3}$$

and established a generalization of the Helly selection principle in which the pointwise limit $f$ of an extracted subsequence belongs to $\mathrm{Reg}(I)$ and which contains all the above selection principles (and many others [13–15]) as particular cases.

In contrast to the previous selection principles, Theorem A involving condition (3) is outside the scope of regulated functions. In fact, if $\mathcal{D} : [0,1] \to \mathbb{R}$ is the Dirichlet function (i.e., $\mathcal{D}(t) = 1$ if $t$ is rational and $\mathcal{D}(t) = 0$ if $t$ is irrational) and $f_j(t) = \mathcal{D}(t)/j$, $t \in [0,1]$, then $f_j \notin \mathrm{Reg}([0,1])$ for all $j \in \mathbb{N}$ and $\nu(n, f_j, [0,1]) = n/j$, and so, condition (3) is satisfied, while the left hand side of (1) is infinite. Moreover, condition (3) is *necessary* for the uniform convergence of $\{f_j\}$ to $f$ if $\nu(n, f, I) = o(n)$, and is "*almost necessary*" for the pointwise convergence (for more details see [12, 13, 15]), which is not at all the case for condition (1).

Let us note that under conditions (1) or (3) a pointwise bounded sequence of functions $\{f_j\}$ is actually uniformly bounded, and that all the above selection principles have been generalized for sequences of functions mapping a nonempty subset of $\mathbb{R}$ into a metric or even uniform space ([1, 5–15] and references therein) and for sequences of monotone functions between linearly ordered sets (Fuchino and Plewik [23]), and some applications of these selection principles to the theory of multifunctions have been given ([2, 10, 17–19]).

Following the philosophy of Theorem A under conditions (1) or (3), which is clearly related to characterizations of regulated functions, we may guess that we will be able to obtain a new pointwise selection principle provided we have some other characterization of regulated functions. Such a characterization, different from all the above, can be given as follows (see Dudley and Norvaiša [22, Part III, Section 2]):

if $\varepsilon > 0$ and $f : I \to \mathbb{R}$, and if we set[a]

$$N(\varepsilon, f, I) = \sup\Big\{n \in \mathbb{N} \;\Big|\; \exists \{I_i\}_1^n \prec I \text{ such that } \min_{1 \le i \le n} |f(I_i)| > \varepsilon\Big\}, \tag{4}$$

then $\mathrm{Reg}(I) = \{f : I \to \mathbb{R} \mid N(\varepsilon, f, I) < \infty \text{ for all } \varepsilon > 0\}$. A preliminary verification of the philosophical idea (Chistyakov and Tretyachenko [20]) leads to Theorem A where condition (1) is replaced by the condition

$$\limsup_{j\to\infty} N(\varepsilon, f_j, I) < \infty \quad \text{for all} \quad \varepsilon > 0 \tag{5}$$

and $f \in \mathrm{BV}_*(I)$ in the conclusion of Theorem A—by $f \in \mathrm{Reg}(I)$. The resulting selection principle is very much alike Theorem A involving condition (3): it can handle sequences of non-regulated functions (if $f_j(t) = \mathcal{D}(t)/j$ as above, we have $N(\varepsilon, f_j, [0,1]) = \infty$ for $0 < \varepsilon < 1/j$ and $N(\varepsilon, f_j, [0,1]) = 0$ for $j \ge 1/\varepsilon$, and so, (5) is satisfied), it implies all selection principles mentioned above, condition (5) is necessary for the uniform convergence and is "almost necessary" for the pointwise convergence (note that a different selection principle employing the quantity (4) is obtained by Dudley and Norvaiša [22, Part III, Proposition 2.8] in the class $\mathrm{Reg}(I)$, and so, is a consequence of Theorem A under condition (5)).

---

[a]The etymology of quantity (4) is not quite clear: for its definition and application the authors of [22, p. 215] refer to Taberski [35]. However, no quantity of the form (4) is considered in [35], and so, for the definition of (4) and its application we refer back to [22].

It is our intention in this paper to generalize Theorem A involving condition (5) for sequences of functions mapping a nonempty subset of $\mathbb{R}$ into a metric space, metric semigroup or reflexive separable Banach space, and clarify the relationship between conditions (3) and (5). We will show that, although the quantities $\{\nu(n,f,I)\}_{n=1}^{\infty}$ and $\{N(\varepsilon,f,I)\}_{\varepsilon>0}$ characterize the set $\mathrm{Reg}(I)$ equally well, they produce different pointwise selection principles: (3) implies (5), but not vice versa (cf. Lemma 3.3(c) in Section 3 and an example preceding Corollary 4.1 in Section 4).

## 2. Main Results

Throughout the paper we assume $T \subset \mathbb{R}$ to be a nonempty subset of the reals $\mathbb{R}$, $(X,d)$ a metric space with metric $d$ and $X^T$ the set of all functions $f: T \to X$ mapping $T$ into $X$. Given a sequence of functions $\{f_j\} \equiv \{f_j\}_{j\in\mathbb{N}}$ in $X^T$ and $f \in X^T$, we write $f_j \to f$ on $T$ to denote the *pointwise* (or *everywhere*) *convergence* of $f_j$ to $f$ as $j \to \infty$, i.e., $\lim_{j\to\infty} d(f_j(t), f(t)) = 0$ for all $t \in T$. A sequence $\{f_j\} \subset X^T$ is said to be *pointwise precompact* (on $T$) provided the sequence $\{f_j(t)\}$ is precompact in $X$ (i.e., its closure in $X$ is compact) for all $t \in T$. Given $f \in X^T$ and $I_i = \{s_i, t_i\} \in \{I_i\}_1^n \prec T$ for some $n \in \mathbb{N}$, we set $|f(I_i)| \equiv |f(I_i)|_d = d(f(s_i), f(t_i))$.

For a number $\varepsilon > 0$ and $f \in X^T$ we define the quantity $N(\varepsilon, f, T)$ valued in $\{0\} \cup \mathbb{N} \cup \{\infty\}$ as the greatest number of different $I_i$'s in $T$ with the property that $|f(I_i)| > \varepsilon$ as in (4) where $I$ is replaced by $T$ and $\sup \varnothing = 0$. Given $\varnothing \neq E \subset T$ and $f \in X^T$, we set $N(\varepsilon, f, E) = N(\varepsilon, f|_E, E)$ where $f|_E : E \to X$, defined by $(f|_E)(t) = f(t)$ for all $t \in E$, is the restriction of the function $f$ to the set $E$.

Our first main result is the following *pointwise selection principle* for metric space valued functions of a real variable in terms of the quantity $N(\varepsilon, f, T)$.

**Theorem 2.1.** *If $\varnothing \neq T \subset \mathbb{R}$, $(X,d)$ is a metric space and $\{f_j\} \subset X^T$ is a pointwise precompact sequence of functions such that*

$$N(\varepsilon) \equiv \limsup_{j\to\infty} N(\varepsilon, f_j, T) < \infty \quad \textit{for all} \quad \varepsilon > 0, \tag{6}$$

*then there exists a subsequence of $\{f_j\}$, which converges pointwise on $T$ to a function $f \in X^T$ satisfying $N(\varepsilon, f, T) \leq N(\varepsilon)$ for all $\varepsilon > 0$.*

In order to formulate our second main result, we need some further definitions.

Let $(X, d, +)$ be a *metric semigroup* (e.g., [10, Section 4]), that is, $(X,d)$ is a metric space with metric $d$, $(X,+)$ is an Abelian semigroup with the addition operation $+$ and $d$ is translation invariant: $d(x+z, y+z) = d(x,y)$ for all $x, y, z \in X$. If $\varnothing \neq T \subset \mathbb{R}$, $f, g \in X^T$ and $\{I_i\}_1^n \prec T$ with $I_i = \{s_i, t_i\}$, we set

$$|f, g(I_i)| \equiv |f, g(I_i)|_d = d(f(t_i) + g(s_i), g(t_i) + f(s_i)), \quad i = 1, \ldots, n, \tag{7}$$

and, given $\varepsilon > 0$, we set

$$N(\varepsilon, f, g, T) = \sup\Big\{n \in \mathbb{N} \,\Big|\, \exists \{I_i\}_1^n \prec T \text{ such that } \min_{1\leq i\leq n} |f, g(I_i)| > \varepsilon\Big\}, \tag{8}$$

provided the set under the supremum sign is nonempty, and $N(\varepsilon, f, g, T) = 0$ otherwise.

Given a double sequence $\{\alpha_{jk} \mid j, k \in \mathbb{N}\} \subset \mathbb{R} \cup \{\infty\}$, we set

$$\limsup_{j,k\to\infty} \alpha_{jk} = \lim_{l\to\infty} \sup\{\alpha_{jk} \mid j \geq l \text{ and } k \geq l\}. \tag{9}$$

Our second main result is the following *pointwise selection principle*:

**Theorem 2.2.** *If $\varnothing \neq T \subset \mathbb{R}$, $(X, d, +)$ is a metric semigroup and $\{f_j\} \subset X^T$ is a pointwise precompact sequence of functions such that*

$$\limsup_{j,k\to\infty} N(\varepsilon, f_j, f_k, T) < \infty \quad \textit{for all} \quad \varepsilon > 0, \tag{10}$$

*then a subsequence of $\{f_j\}$ converges pointwise on $T$ to a function from $X^T$.*

A few remarks are in order. First, the pointwise limits $f$ of extracted subsequences in Theorem 2.2 may not satisfy any regularity conditions such as "$N(\varepsilon, f, T) < \infty$ for all $\varepsilon > 0$" from Theorem 2.1 (e.g., $f_j(t) = (1 + (1/j))\mathcal{D}(t)$, $t \in [0, 1]$, $j \in \mathbb{N}$), and so, Theorem 2.2 may be considered as an "irregular" version of Theorem 2.1. Second, for functions with values in a metric semigroup condition (6) implies condition (10) (but not vice versa [16, Example 4]): it follows from (8) and (4) (and inequality (27) from Section 5) that

$$N(\varepsilon, f_j, f_k, T) \leq N(\varepsilon/2, f_j, T) + N(\varepsilon/2, f_k, T), \quad j, k \in \mathbb{N}, \quad \varepsilon > 0.$$

Third, in the proof of Theorem 2.1 we apply the Helly selection principle, which is inapplicable under the assumptions of Theorem 2.2, and in the proof of the latter theorem we employ the Ramsey theorem from formal logic [31]. Thus, in contrast to Theorem 2.1 our second main result is not equivalent to the Helly theorem. The initial application of Ramsey's theorem in the context of pointwise selection principles goes back to Schrader [33], which later on has been extended by Di Piazza and Maniscalco [21] and Chistyakov and Maniscalco [16]. Our application of Ramsey's theorem and the resulting Theorem 2.2 are quite different from those exposed in [21, 33] and closer to the one elaborated in [16].

The paper is organized as follows. In Section 3 we present main properties of the quantity $N(\varepsilon, f, T)$ needed in the proofs of our results. In Section 4 we prove Theorem 2.1, comment on the "almost necessity" of condition (6) (Theorem 4.1) and establish a variant of Theorem 2.1 for almost everywhere convergence (Theorem 4.2). In Section 5 we prove Theorem 2.2. Finally, in Section 6 we establish variants of Theorems 2.1 and 2.2 for sequences of functions with values in a reflexive separable Banach space, involving weak pointwise and weak almost everywhere convergence (Theorems 6.1 and 6.2).

## 3. Properties of $N(\varepsilon, f, T)$ for Metric Space Valued Functions

The two extreme values of $N(\varepsilon, f, T)$, namely, the value 0 (there is no $I_i$ in $T$ such that $|f(I_i)| > \varepsilon$) and the value $\infty$ (there exists an arbitrary collection of $I_i$'s in $T$

satisfying $|f(I_i)| > \varepsilon$), are characterized by the following conditions, respectively (where 'iff' means, as usual, 'if and only if'):

$$N(\varepsilon, f, T) = 0 \quad \text{iff} \quad |f(I)| \le \varepsilon \quad \text{for all} \quad I \prec T, \text{ and} \tag{11}$$

$$N(\varepsilon, f, T) = \infty \quad \text{iff} \quad \forall n \in \mathbb{N}\ \exists \{I_i\}_1^n \prec T \text{ such that } \min_{1 \le i \le n} |f(I_i)| > \varepsilon. \tag{12}$$

If $N(\varepsilon, f, T) > 0$ (or, equivalently, $N(\varepsilon, f, T) \neq 0$), the set under the supremum sign in (4) with $I = T$ is nonempty, and so, $N(\varepsilon, f, T) \in \mathbb{N} \cup \{\infty\}$, while if $N(\varepsilon, f, T) < \infty$ (or, equivalently, $N(\varepsilon, f, T) \neq \infty$), then the set under the supremum sign in (4) is finite, implying $N(\varepsilon, f, T) \in \{0\} \cup \mathbb{N}$. It follows that if $0 < N(\varepsilon, f, T) < \infty$, then, by virtue of (4), we find that $N(\varepsilon, f, T) \in \mathbb{N}$, and for $n \in \mathbb{N}$ we get:

$$n \le N(\varepsilon, f, T) \quad \text{iff} \quad \exists\ \{I_i\}_1^n \prec T \quad \text{such that} \quad \min_{1 \le i \le n} |f(I_i)| > \varepsilon, \tag{13}$$

and

$$\begin{aligned} n > N(\varepsilon, f, T) \ \text{ iff } \ \forall \{I_i\}_1^n \prec T \ \exists \text{ a permutation } \sigma \in S_n \text{ such that} \\ |f(I_{\sigma(k)})| \le \varepsilon \ \ \forall k = 1, \dots, n - N(\varepsilon, f, T). \end{aligned} \tag{14}$$

Recall that, for $f \in X^T$, the quantity

$$\operatorname{osc}(f, T) = \sup_{I \prec T} |f(I)| = \sup_{s,t \in T} d(f(s), f(t))$$

is said to be the *oscillation of* $f$ *on* $T$ (or the *diameter of the image* $f(T) = \{f(t) \mid t \in T\} \subset X$). Taking into account (11), for any $f \in X^T$ we have:

$$\operatorname{osc}(f, T) = \inf \{\varepsilon > 0 \mid N(\varepsilon, f, T) = 0\} \qquad (\inf \varnothing = \infty).$$

Thus, $f$ is *bounded on* $T$ if and only if $\operatorname{osc}(f, T) < \infty$ (by the definition) or, equivalently, if $N(\varepsilon_0, f, T) = 0$ for some $\varepsilon_0 > 0$, in which case $N(\varepsilon, f, T) = 0$ for all $\varepsilon > \operatorname{osc}(f, T)$ and, moreover, if $\operatorname{osc}(f, T) > 0$, then $N(\varepsilon, f, T) = 0$ for all $\varepsilon \ge \operatorname{osc}(f, T)$ by (11). Note also that $f$ is constant on $T$ if and only if $\operatorname{osc}(f, T) = 0$, which is equivalent to $N(\varepsilon, f, T) = 0$ for all $\varepsilon > 0$. It follows that the quantity

$N(\varepsilon, f, T)$ is *completely characterized* whenever $0 < \varepsilon < \operatorname{osc}(f, T)$.

In order to get a better understanding of the quantity $N(\varepsilon, f, T)$, let us consider its behavior on a bit more general classes of functions of bounded generalized variation than those mentioned in Section 1 and on the class of continuous functions.

Let $\varphi : T \times \mathbb{R}^+ \to \mathbb{R}^+$ be such that the function $\varphi(t, \cdot)$ of the second variable is a $\varphi$-function for each $t \in T$ and $\inf_{t \in T} \varphi(t, u) > 0$ for all $u > 0$. A function $f \in X^T$ is said to be *of bounded generalized* $\varphi$*-variation on* $T$ in the sense of Gniłka [24], in which case we write $f \in \mathrm{BV}_\varphi(T; X)$, provided the $\varphi$*-variation* of $f$ defined by[b]

$$V_\varphi(f, T) = \sup\Big\{\sum_{i=1}^{n} \varphi(t_i, |f(I_i)|) : n \in \mathbb{N},\ \{I_i\}_1^n \prec T \text{ and } t_i \in I_i,\ i = 1, \dots, n\Big\}$$

[b]If $T$ is finite, then the value $n_T = \sup\{n \in \mathbb{N} : \exists \{I_i\}_1^n \prec T\}$ is finite as well. In this case the supremum is taken only over all $n \le n_T$.

is finite. Given $f \in \mathrm{BV}_\varphi(T;X)$, we have (cf. [15, proof of Theorem 9]):

$$\operatorname{osc}(f,T) \le 2\max\{u \in \mathbb{R}^+ : \varphi(t_0,u) \le V_\varphi(f,T)\}, \quad t_0 \in T \text{ is fixed}, \tag{15}$$

$$N(\varepsilon,f,T) \le \frac{V_\varphi(f,T)}{\inf_{t\in T}\varphi(t,\varepsilon)} \quad \text{for} \quad 0<\varepsilon<\operatorname{osc}(f,T). \tag{16}$$

These estimates can be refined for $f \in \mathrm{BV}(T;X)$ (i.e., when $\varphi(t,u)=u$):

$$\operatorname{osc}(f,T) \le V(f,T) \quad \text{and} \quad N(\varepsilon,f,T) \le \frac{V(f,T)}{\varepsilon} \quad \text{for} \quad 0<\varepsilon<\operatorname{osc}(f,T).$$

A sequence $\Phi = \{\varphi_i\}_{i=1}^\infty$ of $\varphi$-functions is said to be a *Schramm sequence* [34] if it satisfies the following two conditions: (i) $\varphi_{i+1}(u) \le \varphi_i(u)$ for all $i \in \mathbb{N}$ and $u \in \mathbb{R}^+$, and (ii) the series $\sum_{i=1}^\infty \varphi_i(u)$ diverges for all $u>0$. If $\Lambda = \{\lambda_i\}_{i=1}^\infty$ is a Waterman sequence and $\varphi_i(u) = u/\lambda_i$, $u \in \mathbb{R}^+$, $i \in \mathbb{N}$, then $\Phi = \{\varphi_i\}_{i=1}^\infty$ is a Schramm sequence. A function $f \in X^T$ is said to be *of $\Phi$-bounded variation on $T$* in the sense of Schramm [34], in which case we write $f \in \Phi\mathrm{BV}(I)$ or $f \in \mathrm{BV}_\Phi(I)$, provided the *$\Phi$-variation* of $f$ given by

$$V_\Phi(f,T) = \sup\Big\{\sum_{i=1}^n \varphi_{\sigma(i)}(|f(I_i)|) : n \in \mathbb{N},\ \{I_i\}_1^n \prec I \text{ and } \sigma \in S_n\Big\}$$

is finite (cf. also the footnote on p. 51). Given $f \in \mathrm{BV}_\Phi(T;X)$, we have (cf. [15, proof of Theorem 10]):

$$\operatorname{osc}(f,T) \le \max\{u \in \mathbb{R}^+ : \varphi_1(u) \le V_\Phi(f,T)\}, \tag{17}$$

$$N(\varepsilon,f,T) \le \max\Big\{n \in \mathbb{N} : \sum_{i=1}^n \varphi_i(\varepsilon) \le V_\Phi(f,T)\Big\}, \quad 0<\varepsilon<\operatorname{osc}(f,T). \tag{18}$$

Given a continuous function $f \in X^{[a,b]}$, the *modulus of continuity* of $f$ is a nondecreasing continuous function $\omega \equiv \omega_f : [0,b-a] \to \mathbb{R}^+$ defined by

$$\omega(\rho) = \sup\{d(f(s),f(t)) : s,\, t \in [a,b] \text{ and } |s-t| \le \rho\} \quad \text{if} \quad 0<\rho\le b-a$$

and $\omega(0) = \lim_{\rho\to+0}\omega(\rho) = 0$. By the Weierstrass theorem, $\operatorname{osc}(f,[a,b]) < \infty$. Setting $\omega_-^{-1}(r) = \min\{\rho \in \mathbb{R}^+ : \omega(\rho) = r\}$ for $r \in [0,\omega(b-a)]$, we have

$$N(\varepsilon,f,[a,b]) \le \frac{b-a}{\omega_-^{-1}(\varepsilon)} \quad \text{if} \quad 0<\varepsilon<\operatorname{osc}(f,[a,b]).$$

The main properties of the quantity (4) for an arbitrary function $f \in X^T$ are gathered in the following

**Lemma 3.1.** *(a) If $0<\varepsilon_1<\varepsilon_2$, then $N(\varepsilon_2,f,T) \le N(\varepsilon_1,f,T)$.*

*(b) If $\varepsilon_0>0$ and $N(\varepsilon_0,f,T)<\infty$, then there exists a $\delta>0$ such that $N(\varepsilon,f,T) = N(\varepsilon_0,f,T)$ for all $\varepsilon_0 \le \varepsilon < \varepsilon_0+\delta$.*

*(c) If $\varnothing \ne E_1 \subset E_2 \subset T$, then $N(\varepsilon,f,E_1) \le N(\varepsilon,f,E_2)$ for all $\varepsilon>0$.*

*(d) If $\{f_j\} \subset X^T$ and $f_j \to f$ on $T$, then*

$$N(\varepsilon,f,T) \le \liminf_{j\to\infty} N(\varepsilon,f_j,T) \quad \textit{for all} \quad \varepsilon>0.$$

*(e) If $s, t \in T$, $s < t$ and $\varepsilon > 0$, then $n_t = N(\varepsilon, f, (-\infty, t] \cap T) < \infty$ if and only if $n_s = N(\varepsilon, f, (-\infty, s] \cap T) < \infty$ and $n_{s,t} = N(\varepsilon, f, [s, t] \cap T) < \infty$, in which case there exists $n_* \in \{0, 1\}$ such that $n_t = n_s + n_{s,t} + n_*$.*

*(f) If $osc(f, T) = \infty$, then $N(\varepsilon, f, T) = \infty$ for all $\varepsilon > 0$, and, as a consequence, if $N(\varepsilon, f, T) < \infty$ for some $\varepsilon > 0$ or, a fortiori, if $N(\varepsilon, f, T) < \infty$ for all $\varepsilon > 0$, then $osc(f, T) < \infty$.*

**Proof.** (a) With no loss of generality we may suppose that $N(\varepsilon_2, f, T) > 0$. If $N(\varepsilon_2, f, T) = \infty$ and $n \in \mathbb{N}$ is arbitrary, or if $N(\varepsilon_2, f, T) < \infty$ and $n = N(\varepsilon_2, f, T)$, then, by virtue of (12) and (13), there exists $\{I_i\}_1^n \prec T$ such that $\min_{1 \le i \le n} |f(I_i)| > \varepsilon_2$, and so, since $\varepsilon_2 > \varepsilon_1$, we have $\min_{1 \le i \le n} |f(I_i)| > \varepsilon_1$, which implies $n \le N(\varepsilon_1, f, T)$ according to (4).

(b) Property (a) yields $N(\varepsilon, f, T) \le N(\varepsilon_0, f, T)$ for all $\varepsilon \ge \varepsilon_0$. It follows that if $N(\varepsilon_0, f, T) = 0$, then $N(\varepsilon, f, T) = 0$ for all $\varepsilon \ge \varepsilon_0$. If $n_0 = N(\varepsilon_0, f, T) \in \mathbb{N}$, then, by (13), there exists $\{I_i\}_1^{n_0} \prec T$ such that $\min_{1 \le i \le n_0} |f(I_i)| > \varepsilon_0$. We set $\delta = \min_{1 \le i \le n_0} |f(I_i)| - \varepsilon_0$. Then $0 < \delta < \infty$, and if $\varepsilon_0 \le \varepsilon < \varepsilon_0 + \delta$, we have $\min_{1 \le i \le n_0} |f(I_i)| > \varepsilon$, and $n_0 \le N(\varepsilon, f, T)$ by (13).

(c) Assume that $N(\varepsilon, f, E_1) > 0$. If $N(\varepsilon, f, E_1) = \infty$ and $n \in \mathbb{N}$ is arbitrary, or if $N(\varepsilon, f, E_1) < \infty$ and $n = N(\varepsilon, f, E_1)$, then, by virtue of (12) and (13), there exists $\{I_i\}_1^n \prec E_1$ such that $\min_{1 \le i \le n} |f(I_i)| > \varepsilon$, and so, since $E_1 \subset E_2$, we have $\{I_i\}_1^n \prec E_2$ and $\min_{1 \le i \le n} |f(I_i)| > \varepsilon$, which gives $n \le N(\varepsilon, f, E_2)$ in accordance with (4).

(d) Suppose $N(\varepsilon, f, T) > 0$, and let us show that if $N(\varepsilon, f, T) = \infty$ and $n \in \mathbb{N}$ is arbitrary, or if $N(\varepsilon, f, T) < \infty$ and $n = N(\varepsilon, f, T)$, then $n \le \liminf_{j \to \infty} N(\varepsilon, f_j, T)$. Applying (12) and (13) we find $\{I_i\}_1^n \prec T$ with $I_i = \{s_i, t_i\}$ such that $\min_{1 \le i \le n} |f(I_i)| > \varepsilon$. Let $\varepsilon' = \varepsilon'(n) > 0$ be such that $\min_{1 \le i \le n} |f(I_i)| > \varepsilon' > \varepsilon$. The pointwise convergence of $f_j$ to $f$ on $T$ implies the existence of a number $J \in \mathbb{N}$ (depending on $\{I_i\}_1^n$) such that, for all $j \ge J$ and $i = 1, \ldots, n$, we have:

$$d(f(s_i), f_j(s_i)) \le \frac{\varepsilon' - \varepsilon}{2} \quad \text{and} \quad d(f_j(t_i), f(t_i)) \le \frac{\varepsilon' - \varepsilon}{2}.$$

By the triangle inequality, for these $j$'s and $i$'s we get:

$$\begin{aligned} \varepsilon' &< |f(I_i)| = d(f(s_i), f(t_i)) \\ &\le d(f(s_i), f_j(s_i)) + d(f_j(s_i), f_j(t_i)) + d(f_j(t_i), f(t_i)) \\ &\le \frac{\varepsilon' - \varepsilon}{2} + |f_j(I_i)| + \frac{\varepsilon' - \varepsilon}{2} = |f_j(I_i)| + \varepsilon' - \varepsilon, \end{aligned} \tag{19}$$

and so, $\min_{1 \le i \le n} |f_j(I_i)| > \varepsilon$ for all $j \ge J$. Definition (4) yields the inequality $n \le N(\varepsilon, f_j, T)$ for all $j \ge J$, whence

$$n \le \inf_{j \ge J} N(\varepsilon, f_j, T) \le \liminf_{j \to \infty} N(\varepsilon, f_j, T).$$

(e) By property (c), $n_s \le n_t$ and $n_{s,t} \le n_t$, and so, if $n_t = 0$, then $n_s = n_{s,t} = 0$. In the rest of this proof we assume that $n_t > 0$.

1. Let us show that if $n_t < \infty$ (in this case $n_s$ and $n_{s,t}$ are finite), then $n_s + n_{s,t} \le n_t$. If $n_s = 0$ or $n_{s,t} = 0$, then $n_s + n_{s,t} \le n_t$, and so, suppose that $n_s > 0$ and $n_{s,t} > 0$. Property (13) implies the existence of $\{I_i\}_1^{n_s} \prec (-\infty, s] \cap T$ and $\{I'_j\}_1^{n_{s,t}} \prec [s,t] \cap T$ such that $\min_{1\le i\le n_s} |f(I_i)| > \varepsilon$ and $\min_{1\le j\le n_{s,t}} |f(I'_j)| > \varepsilon$. Noting that $\{I_i\}_1^{n_s} \cup \{I'_j\}_1^{n_{s,t}} \prec (-\infty, t] \cap T$ and that $|f(I_i)| > \varepsilon$ for all $i = 1, \ldots, n_s$ and $|f(I'_j)| > \varepsilon$ for all $j = 1, \ldots, n_{s,t}$, by virtue of (13) we obtain the inequality $n_s + n_{s,t} \le n_t$.

2. Let $n_s < \infty$ and $n_{s,t} < \infty$. We are going to show that if $n \in \mathbb{N}$, $\{I_i\}_1^n \prec (-\infty, t] \cap T$, where $I_i = [s_i, t_i]$, and $\min_{1\le i\le n} |f(I_i)| > \varepsilon$ (note that such $I_i$'s do exist because $n_t > 0$), then $n \le n_s + n_{s,t} + 1$; the arbitrariness of $n$ and (4) will imply the inequality $n_t \le n_s + n_{s,t} + 1$ and, along with it, the desired equality as well. There are four possibilities: (i) If $\{I_i\}_1^n \prec (-\infty, s] \cap T$, then $n \le n_s$ by virtue of (13), and, analogously, (ii) if $\{I_i\}_1^n \prec [s,t] \cap T$, then $n \le n_{s,t}$. (iii) Suppose that $s$ is in the interior of $I_k$ for some $k \in \{1, \ldots, n\}$. If $n = 1$, then clearly $n \le n_s + n_{s,t} + 1$, and if $n \ge 2$, then we have: $\{I_i\}_1^{k-1} \prec (-\infty, s] \cap T$ and $\{I_i\}_{k+1}^n \prec [s,t] \cap T$, where $\{I_i\}_1^0 = \varnothing = \{I_i\}_{n+1}^n$. It follows from (13) that $k - 1 \le n_s$ and $n - k \le n_{s,t}$, and so, $n \le n_s + n_{s,t} + 1$. (iv) If $n \ge 2$ and $s \in I_k \cap I_{k+1}$ for some $k \in \{1, \ldots, n-1\}$, then $\{I_i\}_1^k \prec (-\infty, s] \cap T$ and $\{I_i\}_{k+1}^n \prec [s,t] \cap T$, and so, $k \le n_s$ and $n - k \le n_{s,t}$ by (13), whence $n \le n_s + n_{s,t}$. Thus, in all the cases (i)–(iv) we have $n \le n_s + n_{s,t} + 1$, which proves (e).

(f) First, we note that if $\operatorname{osc}(f, T) = \infty$, then $\sup_{t\in T} d(f(t), f(s)) = \infty$ for all $s \in T$. Let $\varepsilon > 0$ be arbitrarily fixed. We choose $s_0 \in T$ arbitrarily and, inductively, we construct a sequence $\{s_k\}_{k=0}^\infty \subset T$ as follows: pick $s_1 \in T$ such that $d(f(s_1), f(s_0)) > \varepsilon$, and if $k \ge 2$ and points $s_0, s_1, \ldots, s_{k-1} \in T$ are already chosen, we pick $s_k \in T$ such that

$$d(f(s_k), f(s_{k-1})) > \varepsilon + \sum_{i=1}^{k-1} d(f(s_i), f(s_{i-1})). \tag{20}$$

Now, let $n \in \mathbb{N}$. Denote by $t_0, t_1, \ldots, t_n$ the collection of points $s_0, s_1, \ldots, s_n$ ordered in ascending order, and set $I_i = \{t_{i-1}, t_i\}$, $i = 1, \ldots, n$, so that $\{I_i\}_1^n \prec T$. If we show that $\min_{1\le i\le n} |f(I_i)| > \varepsilon$, then the arbitrariness of $n$ and (12) will yield $N(\varepsilon, f, T) = \infty$. So, let $i \in \{1, \ldots, n\}$ and $I_i = \{s_k, s_m\}$ for some $k, m \in \{0, 1, \ldots, n\}$, $k > m$ (with no loss of generality), so that $|f(I_i)| = d(f(s_k), f(s_m))$. If $m = k - 1$, then $|f(I_i)| > \varepsilon$ by virtue of (20). If $m \le k - 2$, the triangle inequality implies

$$\begin{aligned} d(f(s_k), f(s_{k-1})) &\le d(f(s_k), f(s_m)) + \sum_{i=m+1}^{k-1} d(f(s_{i-1}), f(s_i)) \\ &\le |f(I_i)| + \sum_{i=1}^{k-1} d(f(s_i), f(s_{i-1})), \end{aligned}$$

which together with (20) proves that $|f(I_i)| > \varepsilon$. □

**Remark 3.1.** The case $n_* = 1$ is quite possible in Lemma 3.1(e): if $f : [0,1] \to \mathbb{R}$ is given by $f(t) = 0$ if $0 \le t < 1/2$, $f(1/2) = 1/2$ and $f(t) = 1$ if $1/2 < t \le 1$, and $\varepsilon = 1/2$, then $N(\varepsilon, f, [0,1]) = 1$ and $N(\varepsilon, f, [0,1/2]) = 0 = N(\varepsilon, f, [1/2,1])$.

A function $f \in X^{[a,b]}$ is said to be *regulated* (or *proper*, or *simple*) if it satisfies the Cauchy condition at every point of $[a,b]$, that is, $d(f(s), f(t)) \to 0$ as $s, t \to \tau - 0$ for each $a < \tau \le b$ and $d(f(s), f(t)) \to 0$ as $s, t \to \tau + 0$ for each point $a \le \tau < b$. Note that if $a < \tau \le b$, the condition $\lim_{s,t\to\tau-0} d(f(s), f(t)) = 0$ is equivalent to

$$\forall\, \varepsilon > 0\ \exists\, \delta(\varepsilon) > 0 \text{ such that } |f(I)| \le \varepsilon \text{ for all } I \prec [\tau - \delta(\varepsilon), \tau),$$

and this, by virtue of (11), is equivalent to the following:

$$\forall\, \varepsilon > 0\ \exists\, \delta(\varepsilon) > 0 \text{ such that } N(\varepsilon, f, [\tau - \delta(\varepsilon), \tau)) = 0.$$

A similar remark applies to $a \le \tau < b$ and $\lim_{s,t\to\tau+0} d(f(s), f(t)) = 0$.

If $X$ is a complete metric space, then by virtue of the Cauchy criterion a function $f \in X^{[a,b]}$ is regulated if and only if, at each point $a < \tau \le b$, the left limit $f(\tau - 0) \in X$ exists (and so, $d(f(\tau - 0), f(t)) \to 0$ as $t \to \tau - 0$) and at each point $a \le \tau < b$ the right limit $f(\tau + 0) \in X$ exists (and so, $d(f(t), f(\tau + 0)) \to 0$ as $t \to \tau + 0$).

The following illustrative result generalizes the corresponding assertion from [22, Part III, Theorem 2.1 (I), (II)].

**Lemma 3.2.** *If $(X, d)$ is a metric space, the function $f \in X^{[a,b]}$ is regulated if and only if $N(\varepsilon, f, [a,b]) < \infty$ for all $\varepsilon > 0$.*

**Proof.** *Sufficiency.* Let $a < \tau \le b$ be arbitrary (the arguments for $a \le \tau < b$ are similar). Let us show that for every $\varepsilon > 0$ there exists $\delta(\varepsilon) \in (0, \tau - a)$ such that $d(f(s), f(t)) \le \varepsilon$ for all $s, t \in [\tau - \delta(\varepsilon), \tau)$. On the contrary, suppose that there exists an $\varepsilon_0 > 0$ such that if $\delta \in (0, \tau - a)$, then there are $s_\delta, t_\delta \in [\tau - \delta, \tau)$, $s_\delta < t_\delta$, for which $d(f(s_\delta), f(t_\delta)) > \varepsilon_0$. Given $\delta_1 \in (0, \tau - a)$, pick $s_1, t_1 \in [\tau - \delta_1, \tau)$, $s_1 < t_1$, such that $d(f(s_1), f(t_1)) > \varepsilon_0$. Inductively, if $i \in \mathbb{N}$, $i \ge 2$, and $\delta_{i-1} \in (0, \tau - a)$ and points $s_{i-1}, t_{i-1} \in [\tau - \delta_{i-1}, \tau)$ with $s_{i-1} < t_{i-1}$ are already chosen, we set $\delta_i = \tau - t_{i-1}$ and pick points $s_i, t_i \in [\tau - \delta_i, \tau) = [t_{i-1}, \tau)$, $s_i < t_i$, such that $d(f(s_i), f(t_i)) > \varepsilon_0$. Let $n \in \mathbb{N}$ and $I_i = \{s_i, t_i\}$, $i = 1, \dots, n$. Then, by the construction, $\{I_i\}_1^n \prec (a, \tau) \subset [a,b]$ and $|f(I_i)| > \varepsilon_0$ for all $i = 1, \dots, n$. Since $n$ is arbitrary, (12) implies $N(\varepsilon_0, f, [a,b]) = \infty$, which is a contradiction.

*Necessity.* On the contrary, suppose that $N(\varepsilon_0, f, [a,b]) = \infty$ for some $\varepsilon_0 > 0$. Set $[a_1, b_1] = [a,b]$ and $c_1 = (a_1 + b_1)/2$. By Lemma 3.1(e), $N(\varepsilon_0, f, [a_1, c_1]) = \infty$ or $N(\varepsilon_0, f, [c_1, b_1]) = \infty$. Denote by $[a_2, b_2]$ any of the intervals $[a_1, c_1]$ or $[c_1, b_1]$, for which $N(\varepsilon_0, f, [a_2, b_2]) = \infty$. Inductively, if $k \ge 2$ and the interval $[a_{k-1}, b_{k-1}] \subset [a,b]$ such that $N(\varepsilon_0, f, [a_{k-1}, b_{k-1}]) = \infty$ is already chosen, we set $c_{k-1} = (a_{k-1} + b_{k-1})/2$ and denote by $[a_k, b_k]$ any of the intervals $[a_{k-1}, c_{k-1}]$ or $[c_{k-1}, b_{k-1}]$, for which $N(\varepsilon_0, f, [a_k, b_k]) = \infty$. In this way for each $k \in \mathbb{N}$ we get nested intervals $[a_{k+1}, b_{k+1}] \subset [a_k, b_k]$ from $[a,b]$ such that $N(\varepsilon_0, f, [a_k, b_k]) = \infty$ and $b_k - a_k = (b-a)/2^k$. Let $\tau \in [a,b]$ be the common point of all intervals $[a_k, b_k]$, and so,

$a_k \to \tau$ and $b_k \to \tau$ as $k \to \infty$. Suppose that $a < \tau < b$ (the cases $\tau = a$ or $\tau = b$ are completely similar). Since $a_k \le \tau \le b_k$ and $N(\varepsilon_0, f, [a_k, b_k]) = \infty$ for all $k \in \mathbb{N}$, by virtue of Lemma 3.1(e), there exists a subsequence $\{k_p\}_{p=1}^{\infty}$ of $\{k\}_{k=1}^{\infty}$ such that $N(\varepsilon_0, f, [a_{k_p}, \tau]) = \infty$ for all $p \in \mathbb{N}$ or $N(\varepsilon_0, f, [\tau, b_{k_p}]) = \infty$ for all $p \in \mathbb{N}$. To be more specific, assume that the former possibility takes place; in this case $a_{k_p} \ne \tau$ for all $p \in \mathbb{N}$. For any $\delta \in (0, \tau - a)$ choose a number $p = p(\delta) \in \mathbb{N}$ such that $a_{k_p} \in [\tau - \delta, \tau)$; it follows that $[a_{k_p}, \tau] \subset [\tau - \delta, \tau]$, and so, $N(\varepsilon_0, f, [\tau - \delta, \tau]) = \infty$ according to Lemma 3.1(c). This will imply $N(\varepsilon_0, f, [\tau - \delta, \tau)) = \infty$ (see below), and so, by virtue of (12) there exist $s, t \in [\tau - \delta, \tau)$ such that $d(f(s), f(t)) > \varepsilon_0$, which contradicts the Cauchy condition at $\tau - 0$.

To end the proof, we show that $N(\varepsilon_0, f, [\tau - \delta, \tau)) = \infty$. In fact, the quantity on the left is non-zero, for otherwise, (11) would imply $d(f(s), f(t)) \le \varepsilon_0$ for all $s, t \in [\tau - \delta, \tau)$, and so, $N(\varepsilon_0, f, [\tau - \delta, \tau]) \le 1$. Now if $n \in \mathbb{N}$, then conditions $N(\varepsilon_0, f, [\tau - \delta, \tau]) = \infty$ and (12) yield the existence of $\{I_i\}_1^n \prec [\tau - \delta, \tau]$ such that $|f(I_i)| > \varepsilon_0$ for all $i = 1, \ldots, n$. This in turn gives $\{I_i\}_1^{n-1} \prec [\tau - \delta, \tau)$ and $|f(I_i)| > \varepsilon_0$ for all $i = 1, \ldots, n-1$, which by virtue of (4) means that $n - 1 \le N(\varepsilon_0, f, [\tau - \delta, \tau))$ for arbitrary $n$. $\square$

The *modulus of variation* $\{\nu(n, f, T)\}_{n=1}^{\infty}$ of a function $f \in X^T$ is introduced as in (2) with $I$ replaced by $T$; more precisely, we apply (2) if $n_T = \infty$ (cf. the footnote on p. 51), otherwise if $n_T$ is finite, we apply (2) for $n \le n_T$ and set $\nu(n, f, T) = \nu(n_T, f, T)$ for all $n > n_T$.

The following relationship holds between $\nu(n, f, T)$ and $N(\varepsilon, f, T)$ (in the next lemma items (a) and (b) generalize Theorems 2.1 (II), (III) and 2.2 (a), (b) from [22, Part III] given for $T = [a, b]$ and $X = \mathbb{R}$, and assertion (c) is new).

**Lemma 3.3.** *Suppose $f \in X^T$ and $\{f_j\} \subset X^T$. We have:*

*(a)* $\nu(n, f, T) = o(n)$ *iff* $N(\varepsilon, f, T) < \infty$ *for all* $\varepsilon > 0$;
*(b) given* $j_0 \in \mathbb{N}$, *we have:* $\sup_{j \ge j_0} \nu(n, f_j, T) = o(n)$ *iff* $\sup_{j \ge j_0} N(\varepsilon, f_j, T) < \infty$ *for all* $\varepsilon > 0$ *and* $\sup_{j \ge j_0} \operatorname{osc}(f_j, T) < \infty$;
*(c)* $\limsup_{j \to \infty} \nu(n, f_j, T) = o(n)$ *iff* $\limsup_{j \to \infty} N(\varepsilon, f_j, T) < \infty$ *for all* $\varepsilon > 0$ *and* $\limsup_{j \to \infty} \operatorname{osc}(f_j, T) < \infty$.

**Proof.** (a) follows from (b) if we set $f_j = f$ for all $j \ge j_0$ and note that, by Lemma 3.1(f), condition $N(\varepsilon, f, T) < \infty$ for all $\varepsilon > 0$ implies $\operatorname{osc}(f, T) < \infty$.

(b) *Necessity.* Let $\varepsilon > 0$. With no loss of generality we may assume that $\sup_{j \ge j_0} N(\varepsilon, f_j, T) > 0$. By the assumption, there exists $n_0(\varepsilon) \in \mathbb{N}$ such that $\sup_{j \ge j_0} \nu(n, f_j, T) \le \varepsilon n$ for all $n \ge n_0(\varepsilon)$. We assert that $\sup_{j \ge j_0} N(\varepsilon, f_j, T) \le n_0(\varepsilon)$. In fact, if $j \ge j_0$ is such that $N(\varepsilon, f_j, T) > 0$, and $n \in \mathbb{N}$ is such that there exists $\{I_i\}_1^n \prec T$ with $\min_{1 \le i \le n} |f_j(I_i)| > \varepsilon$, then $n \le n_0(\varepsilon)$, for, otherwise, if $n > n_0(\varepsilon)$, then

$$\sup_{j \ge j_0} \nu(n, f_j, T) \ge \nu(n, f_j, T) \ge \sum_{i=1}^{n} |f_j(I_i)| > \varepsilon n,$$

which leads to a contradiction. On the other hand, if we set $n_0 = n_0(1)$, then $\sup_{j\geq j_0} \nu(n_0, f_j, T) \leq n_0$, and so, $\operatorname{osc}(f_j, T) = \nu(1, f_j, T) \leq \nu(n_0, f_j, T) \leq n_0$ for all $j \geq j_0$, which yields $\sup_{j\geq j_0} \operatorname{osc}(f_j, T) \leq n_0$.

*Sufficiency.* Given $\varepsilon > 0$, we set $N(\varepsilon) = \sup_{j\geq j_0} N(\varepsilon, f_j, T) \in \{0\} \cup \mathbb{N}$. Let us fix $j \geq j_0$ arbitrarily. By virtue of (14), for each $n \geq N(\varepsilon) + 1$ we have: for any $\{I_i\}_1^n \prec T$ there exists a permutation $\sigma \in S_n$ such that $|f_j(I_{\sigma(k)})| \leq \varepsilon$ for all $k = 1, \ldots, n - N(\varepsilon, f_j, T)$. It follows that

$$\begin{aligned}\sum_{i=1}^{n} |f_j(I_i)| &= \sum_{k=n-N(\varepsilon,f_j,T)+1}^{n} |f_j(I_{\sigma(k)})| + \sum_{k=1}^{n-N(\varepsilon,f_j,T)} |f_j(I_{\sigma(k)})| \\ &\leq N(\varepsilon, f_j, T)\operatorname{osc}(f_j, T) + \varepsilon(n - N(\varepsilon, f_j, T)) \\ &\leq N(\varepsilon)M + \varepsilon n,\end{aligned}$$

where $M = \sup_{j\geq j_0} \operatorname{osc}(f_j, T)$ is finite by the assumption. Taking the supremum over all $\{I_i\}_1^n \prec T$, we get $\nu(n, f_j, T) \leq N(\varepsilon)M + \varepsilon n$, and since $j \geq j_0$ is arbitrary, it follows that

$$\sup_{j\geq j_0} \nu(n, f_j, T) \leq N(\varepsilon)M + \varepsilon n \leq 2\varepsilon n \quad \text{if} \quad n \geq N(\varepsilon)M/\varepsilon.$$

Thus, for any $n \geq \max\{N(\varepsilon) + 1, N(\varepsilon)M/\varepsilon\}$ we have $\sup_{j\geq j_0} \nu(n, f_j, T)/n \leq 2\varepsilon$, that is, $\sup_{j\geq j_0} \nu(n, f_j, T) = o(n)$.

(c) *Necessity.* Let $\varepsilon > 0$. By the assumption, there exists $n_0(\varepsilon) \in \mathbb{N}$ such that, taking into account the properties of the limit superior,

$$\lim_{k\to\infty} \sup_{j\geq k} \nu(n_0(\varepsilon), f_j, T) < \varepsilon n_0(\varepsilon),$$

and so, there is $j_0(\varepsilon) \in \mathbb{N}$ such that $\nu(n_0(\varepsilon), f_j, T) < \varepsilon n_0(\varepsilon)$ for all $j \geq j_0(\varepsilon)$. Since

$$\operatorname{osc}(f_j, T) = \nu(1, f_j, T) \leq \nu(n_0(\varepsilon), f_j, T) < \varepsilon n_0(\varepsilon) \quad j \geq j_0(\varepsilon), \tag{21}$$

we have:

$$\limsup_{j\to\infty} \operatorname{osc}(f_j, T) \leq \sup_{j\geq j_0(\varepsilon)} \operatorname{osc}(f_j, T) \leq \varepsilon n_0(\varepsilon) < \infty.$$

By virtue of (21), $f_j$ is a bounded function for each $j \geq j_0(\varepsilon)$, and so, the sequence $\{\nu(n, f_j, T)/n\}_{n=1}^{\infty}$ is nonincreasing for all $j \geq j_0(\varepsilon)$ (cf. [13, remark following Lemma 2], [15, inequality (2)]). It follows that $\nu(n, f_j, T)/n \leq \nu(n_0(\varepsilon), f_j, T)/n_0(\varepsilon) < \varepsilon$ for all $n \geq n_0(\varepsilon)$ and $j \geq j_0(\varepsilon)$, which implies $\sup_{j\geq j_0(\varepsilon)} \nu(n, f_j, T) \leq \varepsilon n$ for all $n \geq n_0(\varepsilon)$. The same arguments as in the proof of the necessity part of (b) with $j_0$ replaced by $j_0(\varepsilon)$ show that

$$\sup_{j\geq j_0(\varepsilon)} N(\varepsilon, f_j, T) \leq n_0(\varepsilon),$$

and this inequality yields

$$\limsup_{j\to\infty} N(\varepsilon, f_j, T) \leq \sup_{j\geq j_0(\varepsilon)} N(\varepsilon, f_j, T) \leq n_0(\varepsilon) < \infty.$$

*Sufficiency.* If we suppose that $\limsup_{j\to\infty} \operatorname{osc}(f_j, T) < M < \infty$, then there exists $j_1 \in \mathbb{N}$ such that $\sup_{j\ge j_1} \operatorname{osc}(f_j, T) < M$. Given $\varepsilon > 0$, condition $\limsup_{j\to\infty} N(\varepsilon, f_j, T) < \infty$ implies the existence of $j_2(\varepsilon) \in \mathbb{N}$ such that $N(\varepsilon) = \sup_{j\ge j_2(\varepsilon)} N(\varepsilon, f_j, T) \in \{0\} \cup \mathbb{N}$. Setting $j_0(\varepsilon) = \max\{j_1, j_2(\varepsilon)\}$ and applying the arguments from the proof of the sufficiency part of (b) for $j \ge j_0(\varepsilon)$, we get $\nu(n, f_j, T) \le N(\varepsilon)M + \varepsilon n$ for all $j \ge j_0(\varepsilon)$, and so, if $n \ge \max\{N(\varepsilon)+1, N(\varepsilon)M/\varepsilon\}$, then

$$\limsup_{j\to\infty} \nu(n, f_j, T) \le \sup_{j\ge j_0(\varepsilon)} \nu(n, f_j, T) \le N(\varepsilon)M + \varepsilon n \le 2\varepsilon n$$

or $\frac{1}{n}\limsup_{j\to\infty} \nu(n, f_j, T) \le 2\varepsilon$, i. e., $\limsup_{j\to\infty} \nu(n, f_j, T) = o(n)$, which was to be proved. □

## 4. Functions with Values in a Metric Space: Proofs

Now we are in a position to prove our first main result.

**Proof of Theorem 2.1.** Denote by $\operatorname{Mon}(T; \mathbb{N})$ the set of all nondecreasing bounded functions mapping $T$ into $\mathbb{N}$. Note that, given $\varepsilon > 0$, by Lemma 3.1(c) the function $t \mapsto N(\varepsilon, f_j, (-\infty, t] \cap T)$ is nondecreasing in $t \in T$ for each $j \in \mathbb{N}$.

1. Making use of the diagonal process we show that, given a decreasing sequence of positive numbers $\{\alpha_k\}_{k=1}^{\infty}$ tending to zero, there exists a subsequence of $\{f_j\}$, again denoted by $\{f_j\}$, and for each $k \in \mathbb{N}$ there exists a function $n_k \in \operatorname{Mon}(T; \mathbb{N})$ such that

$$\lim_{j\to\infty} N(\alpha_k, f_j, (-\infty, t] \cap T) = n_k(t) \quad \text{for all } k \in \mathbb{N} \text{ and } t \in T. \tag{22}$$

By condition (6), for each $\varepsilon > 0$ there are positive integers $K(\varepsilon)$ and $j_0(\varepsilon)$ such that $N(\varepsilon, f_j, T) \le K(\varepsilon)$ for all $j \ge j_0(\varepsilon)$. By virtue of Lemma 3.1(c), the sequence $\{t \mapsto N(\alpha_1, f_j, (-\infty, t] \cap T)\}_{j=j_0(\alpha_1)}^{\infty} \subset \operatorname{Mon}(T; \mathbb{N})$ is uniformly bounded on $T$ by $K(\alpha_1)$, and so, by the Helly selection principle, there exists a subsequence $\{f_{J_1(j)}\}_{j=1}^{\infty}$ of $\{f_j\}_{j=j_0(\alpha_1)}^{\infty}$ (where $J_1 : \mathbb{N} \to \mathbb{N}$ is a strictly increasing subsequence of $\{j_0(\alpha_1)+j-1\}_{j=1}^{\infty}$) and a function $n_1 \in \operatorname{Mon}(T; \mathbb{N})$ such that $N(\alpha_1, f_{J_1(j)}, (-\infty, t] \cap T)$ converges to $n_1(t)$ as $j \to \infty$ for all $t \in T$. Choose the least number $j_1 \in \mathbb{N}$ such that $J_1(j_1) \ge j_0(\alpha_2)$. Inductively, if $k \ge 2$ and a subsequence $\{f_{J_{k-1}(j)}\}_{j=1}^{\infty}$ of the initial sequence $\{f_j\}$ and a number $j_{k-1} \in \mathbb{N}$ such that $J_{k-1}(j_{k-1}) \ge j_0(\alpha_k)$ are already chosen, by Helly's selection principle applied to the sequence of functions

$$\{t \mapsto N(\alpha_k, f_{J_{k-1}(j)}, (-\infty, t] \cap T)\}_{j=j_{k-1}}^{\infty} \subset \operatorname{Mon}(T; \mathbb{N}),$$

which is uniformly bounded by $K(\alpha_k)$, we find a subsequence $\{f_{J_k(j)}\}_{j=1}^{\infty}$ of $\{f_{J_{k-1}(j)}\}_{j=j_{k-1}}^{\infty}$ and a function $n_k \in \operatorname{Mon}(T; \mathbb{N})$ such that

$$\lim_{j\to\infty} N(\alpha_k, f_{J_k(j)}, (-\infty, t] \cap T) = n_k(t) \quad \text{for all } \ t \in T.$$

It follows that the diagonal sequence $\{f_{J_j(j)}\}_{j=1}^{\infty}$, again denoted by $\{f_j\}$, satisfies condition (22).

2. Let $Q$ denote an at most countable dense subset of $T$ (implying $Q \subset T \subset$ closure of $Q$). Note that any point $t \in T$, which is not a limit point for $T$, belongs to $Q$. Since $n_k \in \mathrm{Mon}\,(T;\mathbb{N})$, the set $Q_k \subset T$ of its points of discontinuity is at most countable for all $k \in \mathbb{N}$. Setting $S = Q \cup \bigcup_{k=1}^{\infty} Q_k$, we find that $S$ is an at most countable dense subset of $T$ and, moreover, if $T \setminus S \neq \varnothing$, then

$$\text{each function } n_k \text{ is continuous at all points of } T \setminus S,\ k \in \mathbb{N}. \tag{23}$$

Since the set $\{f_j(t)\}$ is precompact in $X$ for all $t \in T$ and $S \subset T$ is at most countable, with no loss of generality we may assume (again applying the diagonal process and passing to a subsequence of $\{f_j\}$ if necessary) that $f_j(s)$ converges in $X$ as $j \to \infty$ to a point denoted by $f(s) \in X$ for all $s \in S$. If $T = S$, we are through.

3. Now suppose $T \neq S$. Let us prove that, given $t \in T \setminus S$, the sequence $\{f_j(t)\}$ converges in $X$. For this, let $\varepsilon > 0$ be arbitrarily fixed. Since the sequence $\{\alpha_k\}$ from step 1 tends to zero, we choose and fix $k = k(\varepsilon) \in \mathbb{N}$ such that $\alpha_k \le \varepsilon$. The definition of $S$ and (23) imply that the point $t$ is a limit point for $T$ and a point of continuity of $n_k$, so by the density of $S$ in $T$ there exists $s = s(k,t) \in S$ such that $|n_k(t) - n_k(s)| < 1$ or, equivalently, $n_k(t) = n_k(s)$. Property (22) yields the existence of positive integers $j_1 = j_1(k,t)$ and $j_2 = j_2(k,s)$ such that if $j \ge \max\{j_1, j_2\}$, then

$$N(\alpha_k, f_j, (-\infty, t] \cap T) = n_k(t) \quad \text{and} \quad N(\alpha_k, f_j, (-\infty, s] \cap T) = n_k(s).$$

Assuming with no loss of generality that $s < t$ and applying Lemma 3.1(e), we get, for all $j \ge \max\{j_1, j_2\}$:

$$\begin{aligned} N(\alpha_k, f_j, [s,t] \cap T) &\le N(\alpha_k, f_j, (-\infty, t] \cap T) - N(\alpha_k, f_j, (-\infty, s] \cap T) \\ &= n_k(t) - n_k(s) = 0. \end{aligned}$$

It follows that $N(\alpha_k, f_j, [s,t] \cap T) = 0$, and so, by virtue of (11) we have, in particular, $d(f_j(s), f_j(t)) \le \alpha_k \le \varepsilon$ for all $j \ge \max\{j_1, j_2\}$. Since $\{f_j(s)\}$ is convergent, it is a Cauchy sequence, and so, there exists a positive integer $j_3 = j_3(\varepsilon, s)$ such that $d(f_j(s), f_{j'}(s)) \le \varepsilon$ for all $j, j' \ge j_3$. Hence, the number $j_4 = \max\{j_1, j_2, j_3\}$ depends on $\varepsilon$ only and, for all $j, j' \ge j_4$, we have:

$$d(f_j(t), f_{j'}(t)) \le d(f_j(t), f_j(s)) + d(f_j(s), f_{j'}(s)) + d(f_{j'}(s), f_{j'}(t)) \le 3\varepsilon.$$

Thus, $\{f_j(t)\}$ is a Cauchy sequence in $X$ and, since it is precompact in $X$, it is convergent in $X$ to a point denoted by $f(t) \in X$.

4. The function $f \in X^T$ defined at the end of steps 2 and 3 is the pointwise limit on $T$ of the sequence $\{f_j\}$ (which is a subsequence of the original sequence). Applying Lemma 3.1(d), we conclude that

$$N(\varepsilon, f, T) \le \liminf_{j \to \infty} N(\varepsilon, f_j, T) \le \limsup_{j \to \infty} N(\varepsilon, f_j, T) \le N(\varepsilon) \quad \forall\, \varepsilon > 0. \qquad \square$$

If condition (6) is not valid, Theorem 2.1 is wrong: no subsequence of $f_j(t) = \sin(jt)$, $t \in [0, 2\pi]$, $j \in \mathbb{N}$, is pointwise convergent on $[0, 2\pi]$ and

$$4j \le N(\varepsilon, f_j, [0, 2\pi]) \le 4j/\varepsilon \quad \text{for all} \quad 0 < \varepsilon < 1 \quad \text{and} \quad j \in \mathbb{N}$$

(more details can be found in [13, example 3 in Section 3], [20, Section 3.2]).

It is to be noted that we did not suppose for $j \in \mathbb{N}$ that $N(\varepsilon, f_j, T)$ is finite for all $\varepsilon > 0$ in Theorem 2.1, however, we always have the "regularity" condition that $N(\varepsilon, f, T)$ is finite for all $\varepsilon > 0$ for the limit function $f$ (see the example of $f_j(t) = \mathcal{D}(t)/j$ following conditions (3) and (5)). We also note that, by Lemma 3.3(c), condition $\limsup_{j\to\infty} \nu(n, f_j, T) = o(n)$ implies condition (6), and so, the selection principle from Chistyakov [13, Theorem 1] is a consequence of our Theorem 2.1, which allows sequences $\{f_j\}$ with $\limsup_{j\to\infty} \operatorname{osc}(f_j, T) = \infty$. In fact, if we define $f_j : [0, 1] \to \mathbb{R}$ by $f_j(t) = 0$ for $t \neq 1 - (1/j)$ and $f_j(t) = j$ for $t = 1 - (1/j)$, $j \in \mathbb{N}$, then $1 \leq N(\varepsilon, f_j, [0, 1]) \leq 2$ for $0 < \varepsilon < j = \operatorname{osc}(f_j, [0, 1])$ and $j \in \mathbb{N}$.

Applying Theorem 2.1 and the diagonal process we get the following

**Corollary 4.1.** *If a pointwise precompact sequence of functions $\{f_j\} \subset X^T$ satisfies condition $\limsup_{j\to\infty} N(\varepsilon, f_j, T \cap [s, t]) < \infty$ for all $[s, t] \prec T$ and $\varepsilon > 0$ or condition $\limsup_{j\to\infty} N(\varepsilon, f_j, T \setminus E) < \infty$ for some at most countable $E \subset T$ and all $\varepsilon > 0$, then it contains a subsequence which converges pointwise on $T$ to a function $f \in X^T$ such that $N(\varepsilon, f, T \cap [s, t]) < \infty$ for all $[s, t] \prec T$ and $\varepsilon > 0$ or, respectively, $N(\varepsilon, f, T \setminus E) < \infty$ for all $\varepsilon > 0$.*

Two more modes of convergence of $\{f_j\} \subset X^T$ to $f \in X^T$ will be of significance: the *uniform convergence*, i.e., $\lim_{j\to\infty} \sup_{t\in T} d(f_j(t), f(t)) = 0$, written as $f_j \rightrightarrows f$ on $T$, and the *almost everywhere convergence*, i.e., $f_j \to f$ on $T \setminus E$ for some set $E \subset T$ of Lebesgue measure zero, written as $f_j \to f$ a.e. on $T$.

**Theorem 4.1.** *Let $\varnothing \neq T \subset \mathbb{R}$, $(X, d)$ be a metric space, $\{f_j\} \subset X^T$ and $f \in X^T$. First suppose that $f_j \rightrightarrows f$ on $T$. Then we have:*

*(a) $\limsup\limits_{j\to\infty} N(\varepsilon, f_j, T) \leq \lim\limits_{\delta\to\varepsilon-0} N(\delta, f, T)$ for all $\varepsilon > 0$, and the right hand side in this inequality cannot be replaced by $N(\varepsilon, f, T)$;*

*(b) if $N(\varepsilon, f, T) < \infty$ for all $\varepsilon > 0$, then $\limsup_{j\to\infty} N(\varepsilon, f_j, T) < \infty$ for all $\varepsilon > 0$; however, it may happen that condition "$N(\varepsilon, f_j, T) < \infty$ for all $\varepsilon > 0$" does not hold for any $j \in \mathbb{N}$;*

*(c) if $N(\varepsilon, f_j, T) < \infty$ for all $\varepsilon > 0$ and $j \in \mathbb{N}$, then $N(\varepsilon, f, T) < \infty$ for all $\varepsilon > 0$.*

*Assertions (a)–(c) are false for pointwise convergence $f_j \to f$ on $T$.*

*(d) Suppose $T$ is a measurable set with finite Lebesgue measure, $\{f_j\} \subset X^T$ is a sequence of measurable functions, $f \in X^T$, $N(\varepsilon, f, T) < \infty$ for all $\varepsilon > 0$ and $f_j \to f$ pointwise or a.e. on $T$. Then for each $\eta > 0$ there exists a measurable set $E = E(\eta) \subset T$ of measure $\leq \eta$ such that $\limsup_{j\to\infty} N(\varepsilon, f_j, T \setminus E) < \infty$ for all $\varepsilon > 0$.*

**Proof.** (a) Let $\varepsilon > 0$ and $0 < \delta < \varepsilon$. Since $f_j \rightrightarrows f$ on $T$, there exists a number $J = J(\varepsilon, \delta) \in \mathbb{N}$ such that $\sup_{t\in T} d(f_j(t), f(t)) \leq (\varepsilon - \delta)/2$ for all $j \geq J$. Suppose that we have already shown that

$$N(\varepsilon, f_j, T) \leq N(\delta, f, T) \quad \text{for all} \quad j \geq J; \tag{24}$$

then this will imply

$$\limsup_{j\to\infty} N(\varepsilon, f_j, T) \leq \sup_{j\geq J} N(\varepsilon, f_j, T) \leq N(\delta, f, T) \quad \text{for all } \delta \in (0, \varepsilon),$$

and it remains to take into account the fact that, by Lemma 3.1(a), the function $\delta \mapsto N(\delta, f, T)$ is nonincreasing and pass to the limit as $\delta \to \varepsilon - 0$. Now, in order to obtain (24), with no loss of generality we assume that, given $j \geq J$, we have $N(\varepsilon, f_j, T) > 0$. Let $n \in \mathbb{N}$ be such that there exists $\{I_i\}_1^n \prec T$ with $I_i = \{s_i, t_i\}$ such that $|f_j(I_i)| = d(f_j(s_i), f_j(t_i)) > \varepsilon$ for all $i = 1, \ldots, n$. It follows that

$$\begin{aligned}\varepsilon < d(f_j(s_i), f_j(t_i)) &\leq d(f_j(s_i), f(s_i)) + d(f(s_i), f(t_i)) + d(f(t_i), f_j(t_i)) \\ &\leq d(f(s_i), f(t_i)) + 2\sup_{t\in T} d(f_j(t), f(t)) \leq |f(I_i)| + \varepsilon - \delta,\end{aligned}$$

and so, $|f(I_i)| > \delta$ for all $i = 1, \ldots, n$. Then (4) gives $n \leq N(\delta, f, T)$, and the arbitrariness of $n$ implies (24).

As for the second assertion in (a), consider the sequence $f_j : T = [0,1] \to \mathbb{R}$ given for $j \in \mathbb{N}$ by $f_j(0) = 1 + (1/j)$, $f_j(t) = 0$ if $0 < t < 1$ and $f_j(1) = 2 + (1/j)$. If $f$ is the limit function of $\{f_j\}$, then $N(1, f_j, T) = 2$, $N(1, f, T) = 1$, $N(\delta, f, T) = 2$ for $0 < \delta < 1$, and $N(2, f_j, T) = 1$, $N(2, f, T) = 0$, $N(\delta, f, T) = 1$ for $1 < \delta < 2$.

(b) is an immediate consequence of (a). The example of $f_j(t) = \mathcal{D}(t)/j$ on $[0,1]$ following (5) justifies the second part of (b).

(c) This part resembles the proof of Lemma 3.1(d). On the contrary, suppose that $N(2\varepsilon, f, T) = \infty$ for some $\varepsilon > 0$. Then, on the one hand, there exists $j \in \mathbb{N}$ depending only on $\varepsilon$ such that $\sup_{t\in T} d(f_j(t), f(t)) \leq \varepsilon/2$ and, on the other hand, by (12) for each $n \in \mathbb{N}$ one finds $\{I_i\}_1^n \prec T$ with $I_i = \{s_i, t_i\}$ such that $\min_{1\leq i\leq n} |f(I_i)| > 2\varepsilon$. Now inequality (19) with $\varepsilon' = 2\varepsilon$ yields

$$2\varepsilon < |f(I_i)| \leq |f_j(I_i)| + 2\sup_{t\in T} d(f_j(t), f(t)) \leq |f_j(I_i)| + \varepsilon, \quad i = 1, \ldots, n,$$

and so, $\min_{1\leq i\leq n} |f_j(I_i)| > \varepsilon$ or, due to the arbitrariness of $n$, $N(\varepsilon, f_j, T) = \infty$, which contradicts the assumption.

In order to see that assertions (a)–(c) are wrong for the pointwise convergence, cf. the example of a sequence $\{f_j\}$ preceding Corollary 4.1 for (a) and examples 4 and 5 from [13, Section 3] for (b) and (c).

(d) By the assumptions and Egorov's Theorem, for each $\eta > 0$ there exists a measurable set $E = E(\eta) \subset T$ of measure $\leq \eta$ such that $f_j \rightrightarrows f$ on $T \setminus E$. Since $N(\varepsilon, f, T) < \infty$ for all $\varepsilon > 0$, Lemma 3.1(c) implies $N(\varepsilon, f, T \setminus E) < \infty$ for all $\varepsilon > 0$. Now our assertion follows from Theorem 4.1(b). □

By Theorem 2.1, a pointwise precompact sequence $\{f_j\} \subset X^T$, satisfying

$$\limsup_{j\to\infty} N(\varepsilon, f_j, T \setminus E) < \infty \text{ for all } \varepsilon > 0 \text{ and some } E \subset T \text{ of measure zero,}$$

contains a subsequence which converges a. e. on $T$ to a function $f \in X^T$ such that $N(\varepsilon, f, T \setminus E) < \infty$ for all $\varepsilon > 0$. The following theorem, a *selection principle for the a. e. convergence*, generalizes Theorem 6 from [13] and is a subsequence-converse to Theorem 4.1(d).

**Theorem 4.2.** *Suppose $\varnothing \neq T \subset \mathbb{R}$, $(X, d)$ is a metric space and $\{f_j\} \subset X^T$ is a pointwise precompact (or a. e. precompact) sequence of functions on $T$ satisfying the condition: for each $k \in \mathbb{N}$ there exists a measurable set $E_k \subset T$ of measure $\leq 1/k$ such that*

$$\limsup_{j \to \infty} N(\varepsilon, f_j, T \setminus E_k) < \infty \quad \textit{for all} \quad \varepsilon > 0. \tag{25}$$

*Then a subsequence of $\{f_j\}$ converges a. e. on $T$ to a function $f \in X^T$ having the property: given $k \in \mathbb{N}$, there is a measurable set $E'_k \subset T$ of measure $\leq 1/k$ such that $N(\varepsilon, f, T \setminus E'_k) < \infty$ for all $\varepsilon > 0$.*

**Proof.** We follow the lines of the proof of Theorem 6 from [13] with suitable modifications. Let $T_0 \subset T$ be a set of measure zero such that $\{f_j(t)\}$ is precompact in $X$ for all $t \in T \setminus T_0$. We will apply Theorem 2.1 and the diagonal process. By the assumption, there exists a measurable set $E_1 \subset T$ of measure $\leq 1$, for which (25) holds with $k = 1$. The sequence $\{f_j\}$ is pointwise precompact on $T \setminus (T_0 \cup E_1)$ and, by Lemma 3.1(c), $\limsup_{j \to \infty} N(\varepsilon, f_j, T \setminus (T_0 \cup E_1)) < \infty$ for all $\varepsilon > 0$. Theorem 2.1 implies the existence of a subsequence $\{f_{J_1(j)}\}_{j=1}^{\infty}$ of $\{f_j\}$ and a function $f^{(1)} : T \setminus (T_0 \cup E_1) \to X$ satisfying $N(\varepsilon, f^{(1)}, T \setminus (T_0 \cup E_1)) < \infty$ for all $\varepsilon > 0$ such that $f_{J_1(j)} \to f^{(1)}$ on $T \setminus (T_0 \cup E_1)$ as $j \to \infty$. Assume that $k \geq 2$ and a subsequence $\{f_{J_{k-1}(j)}\}_{j=1}^{\infty}$ of $\{f_j\}$ is already chosen. By (25) and the arguments as above (for $E_1$), the sequence $\{f_{J_{k-1}(j)}\}_{j=1}^{\infty}$ is pointwise precompact on $T \setminus (T_0 \cup E_k)$ and $\limsup_{j \to \infty} N(\varepsilon, f_{J_{k-1}(j)}, T \setminus (T_0 \cup E_k)) < \infty$ for all $\varepsilon > 0$, and so, by Theorem 2.1, we find a subsequence $\{f_{J_k(j)}\}_{j=1}^{\infty}$ of $\{f_{J_{k-1}(j)}\}_{j=1}^{\infty}$ and a function $f^{(k)} : T \setminus (T_0 \cup E_k) \to X$ such that $N(\varepsilon, f^{(k)}, T \setminus (T_0 \cup E_k)) < \infty$ for all $\varepsilon > 0$ and $f_{J_k(j)} \to f^{(k)}$ on $T \setminus (T_0 \cup E_k)$ as $j \to \infty$. The set $E = T_0 \cup \bigcap_{k=1}^{\infty} E_k$ is of measure zero and $T \setminus E = \bigcup_{k=1}^{\infty} (T \setminus (T_0 \cup E_k))$. We define the function $f : T \setminus E \to X$ according to the following rule: given $t \in T \setminus E$, there is $k \in \mathbb{N}$ such that $t \in T \setminus (T_0 \cup E_k)$, and we set $f(t) = f^{(k)}(t)$. By the construction, this definition is correct: the value $f(t)$ is independent of $k$. Then the diagonal sequence $\{f_{J_j(j)}\}_{j=1}^{\infty}$ converges to $f$ pointwise on $T \setminus E$: indeed, each $t \in T \setminus E$ is in $T \setminus (T_0 \cup E_k)$ for some $k \in \mathbb{N}$ implying $f(t) = f^{(k)}(t)$ and, since $\{f_{J_j(j)}\}_{j=k}^{\infty}$ is a subsequence of $\{f_{J_k(j)}\}_{j=1}^{\infty}$, we have: $f_{J_j(j)}(t) \to f^{(k)}(t) = f(t)$ in $X$ as $j \to \infty$. Extending $f$ from $T \setminus E$ to the whole $T$ arbitrarily, denoting this extension again by $f$ and setting $E'_k = T_0 \cup E_k$ for $k \in \mathbb{N}$, we find that $E'_k$ is of measure $\leq 1/k$, $f = f^{(k)}$ on $T \setminus E'_k$ and $N(\varepsilon, f, T \setminus E'_k) = N(\varepsilon, f^{(k)}, T \setminus (T_0 \cup E_k)) < \infty$ for all $\varepsilon > 0$.

□

## 5. Functions with Values in a Metric Semigroup

Throughout this section $(X, d, +)$ designates a metric semigroup. For all elements $x, y, u, v$ of $X$ the following two inequalities hold:

$$d(x, y) \le d(x + u, y + v) + d(u, v), \tag{26}$$

$$d(x + u, y + v) \le d(x, y) + d(u, v), \tag{27}$$

and, in particular, (27) implies that the addition operation $(x, y) \mapsto x + y$ is a continuous mapping from $X \times X$ into $X$.

If $f, g \in X^T$, the quantity $N(\varepsilon, f, g, T)$, introduced in Section 2, takes its values in $\{0\} \cup \mathbb{N} \cup \{\infty\}$ and, because its definition implies relations (11)–(14) where $f$ is replaced by the pair $f, g$, has many properties which are parallel to those of $N(\varepsilon, f, T)$ from Section 3. Note that quantity (8) is symmetric with respect to $f$ and $g$, it is equal to zero if $f = g$, and if $(X, \|\cdot\|)$ is a normed linear space with the induced metric $d(x, y) = \|x - y\|$, $x, y \in X$, then, since (7) is equal to $|(f - g)(I_i)|$, we simply have $N(\varepsilon, f, g, T) = N(\varepsilon, f - g, T)$.

Moreover, assertions (a), (b), (c) and (e) of Lemma 3.1 hold if we replace $N(\varepsilon, f, T)$ in that lemma by $N(\varepsilon, f, g, T)$; below in this section we will refer to this modification of Lemma 3.1 again as Lemma 3.1. In order to prove Theorem 2.2, we also need Ramsey's theorem [31, Theorem A], which we recall below as Theorem B.

Given a nonempty set $\Gamma$, $n \in \mathbb{N}$ and an injective function $\gamma : \{1, \ldots, n\} \to \Gamma$, the set $\{\gamma(1), \ldots, \gamma(n)\}$ is called an *n-combination* of elements of $\Gamma$ (note that an $n$-combination may be generated by $n!$ different injective functions). Let $\Gamma[n]$ denote the family of all $n$-combinations of elements of $\Gamma$.

**Theorem B.** *Suppose $\Gamma$ is an infinite set, $n, m \in \mathbb{N}$, and $\Gamma[n] = \bigcup_{i=1}^{m} C_i$ is a disjoint union of its $m$ nonempty subsets $C_i$. Then, under the Axiom of Choice, $\Gamma$ contains an infinite subset $\Delta$ such that $\Delta[n] \subset C_{i_0}$ for some $i_0 \in \{1, \ldots, m\}$.*

This theorem will be applied several times in the proof of Theorem 2.2 with $\Gamma$ a subset of $\mathbb{N} \times \mathbb{N}$ and $n = 2$. Also, we will make use of the following definition: a double sequence $\{\alpha_{jk} \mid j, k \in \mathbb{N}\}$, having the property that $\alpha_{jj} = 0$ for all $j \in \mathbb{N}$ (cf. (29) and below), is said to *converge* to a number $\ell \in \mathbb{R}$, written as $\lim_{j,k\to\infty} \alpha_{jk} = \ell$, if for each $\varepsilon > 0$ there exists a $j_0(\varepsilon) \in \mathbb{N}$ such that $\alpha_{jk} \in [\ell - \varepsilon, \ell + \varepsilon]$ for all $j \ge j_0(\varepsilon)$ and $k \ge j_0(\varepsilon)$ with $j \ne k$.

Now we are in a position to prove our second main result.

**Proof of Theorem 2.2.** If the domain $T$ is at most countable, then we may apply the precompactness of $\{f_j(t)\}$ for all $t \in T$ and the standard diagonal process, or if there is an increasing sequence $\{k_j\}_{j=1}^{\infty}$ of positive integers such that $f_{k_i} = f_{k_j}$ on $T$ for all $i, j \in \mathbb{N}$, then we may choose the constant subsequence $\{f_{k_j}\}_{j=1}^{\infty}$ of $\{f_j\}$. In both these cases we are done. Thus, in what follows we assume that $T$ is *uncountable* and (picking a subsequence of $\{f_j\}$ if necessary) that all functions in $\{f_j\}$ are *distinct.*

Given $\varepsilon > 0$, we set (cf. (10))

$$M(\varepsilon) = 1 + \limsup_{j,k\to\infty} N(\varepsilon, f_j, f_k, T) \in \mathbb{N}. \tag{28}$$

We divide the rest of the proof into three steps for clarity.

1. Let us prove that, given $\varepsilon > 0$, there exists a subsequence $\{f_j^\varepsilon\}_{j=1}^\infty$ of $\{f_j\}$ and a nondecreasing function $N_\varepsilon : T \to \{0, 1, \ldots, M(\varepsilon)\}$ such that

$$\lim_{j,k\to\infty} N(\varepsilon, f_j^\varepsilon, f_k^\varepsilon, (-\infty, t] \cap T) = N_\varepsilon(t) \quad \text{for all} \quad t \in T. \tag{29}$$

Let us fix $\varepsilon > 0$. By virtue of (28), there exists a $j_0(\varepsilon) \in \mathbb{N}$ such that $N(\varepsilon, f_j, f_k, T) \le M(\varepsilon)$ for all $j, k \in F_\varepsilon$, where $F_\varepsilon = \{j \in \mathbb{N} \mid j \ge j_0(\varepsilon)\}$. If $t \in T$ and $i \in \{0, 1, \ldots, M(\varepsilon)\}$, we set

$$C_i^{\varepsilon,t} = \{(j,k) \in F_\varepsilon \times F_\varepsilon \mid j \ne k \text{ and } N(\varepsilon, f_j, f_k, (-\infty, t] \cap T) = i\}. \tag{30}$$

Putting $\Gamma = F_\varepsilon \times F_\varepsilon$, we get $\Gamma[2] = \bigcup_{i=0}^{M(\varepsilon)} C_i^{\varepsilon,t}$, the sets under the union sign being disjoint, and so, Theorem B implies the existence of an $i(\varepsilon, t) \in \{0, 1, \ldots, M(\varepsilon)\}$ and a strictly increasing sequence $J_{\varepsilon,t} : \mathbb{N} \to F_\varepsilon$ such that

$$N(\varepsilon, f_{J_{\varepsilon,t}(j)}, f_{J_{\varepsilon,t}(k)}, (-\infty, t] \cap T) = i(\varepsilon, t) \quad \text{for all} \quad j, k \in \mathbb{N}. \tag{31}$$

The sets $L_T = \{t \in T \mid (t - \delta, t) \cap T = \varnothing \text{ for some } \delta > 0\}$ of points isolated from the left for $T$ and $R_T = \{t \in T \mid (t, t + \delta) \cap T = \varnothing \text{ for some } \delta > 0\}$ of points isolated from the right for $T$ are at most countable (possibly empty), and so, if $Q$ is an at most countable dense subset of $T$ (i.e., $Q \subset T \subset$ closure of $Q$), then $L_T \cap R_T \subset Q$ and $Z = Q \cup L_T \cup R_T$ is an at most countable dense subset of $T$. With no loss of generality we may assume that $Z = \{s_p\}_{p=1}^\infty$.

By virtue of (31), $i_1(\varepsilon) = i(\varepsilon, s_1) \in \{0, 1, \ldots, M(\varepsilon)\}$, $J_{\varepsilon,1} = J_{\varepsilon,s_1} : \mathbb{N} \to F_\varepsilon$ is strictly increasing and $N(\varepsilon, f_{J_{\varepsilon,1}(j)}, f_{J_{\varepsilon,1}(k)}, (-\infty, s_1] \cap T) = i_1(\varepsilon)$ for all $j, k \in \mathbb{N}$. Inductively, if $p \ge 2$, $i_{p-1}(\varepsilon) \in \{0, 1, \ldots, M(\varepsilon)\}$ and a strictly increasing sequence $J_{\varepsilon,p-1} : \mathbb{N} \to F_\varepsilon$ satisfying

$$N(\varepsilon, f_{J_{\varepsilon,p-1}(j)}, f_{J_{\varepsilon,p-1}(k)}, (-\infty, s_{p-1}] \cap T) = i_{p-1}(\varepsilon) \quad \text{for all} \quad j, k \in \mathbb{N}$$

are already chosen, we apply the arguments (30)–(31) with $t = s_p$ and the pair $f_j, f_k$ replaced by the pair $f_{J_{\varepsilon,p-1}(j)}, f_{J_{\varepsilon,p-1}(k)}$: there exist an $i_p(\varepsilon) \in \{0, 1, \ldots, M(\varepsilon)\}$ and a strictly increasing subsequence $\{J_{\varepsilon,p}(j)\}_{j=1}^\infty \subset F_\varepsilon$ of $\{J_{\varepsilon,p-1}(j)\}_{j=1}^\infty$ such that

$$N(\varepsilon, f_{J_{\varepsilon,p}(j)}, f_{J_{\varepsilon,p}(k)}, (-\infty, s_p] \cap T) = i_p(\varepsilon) \quad \text{for all} \quad j, k \in \mathbb{N}.$$

Then the diagonal subsequence $\{f_{J_{\varepsilon,j}(j)}\}_{j=1}^\infty$ of $\{f_j\}$, denoted by $\{f_j^\varepsilon\} = \{f_j^\varepsilon\}_{j=1}^\infty$, satisfies the condition:

$$\lim_{j,k\to\infty} N(\varepsilon, f_j^\varepsilon, f_k^\varepsilon, (-\infty, s] \cap T) = \psi_\varepsilon(s) \quad \text{for all} \quad s \in Z, \tag{32}$$

where the function $\psi_\varepsilon : Z \to \{0, 1, \ldots, M(\varepsilon)\}$ is defined by $\psi_\varepsilon(s_p) = i_p(\varepsilon)$ for all $p \in \mathbb{N}$. It is clear from (the modification of) Lemma 3.1(c) that $\psi_\varepsilon$ is a nondecreasing function on $Z$.

Now we extend the function $\psi_\varepsilon$ from $Z$ to the whole $\mathbb{R}$ following Saks [32, Chapter 7, Section 4, Lemma (4.1)]: $\widetilde{\psi}_\varepsilon(t) = \sup_{s\in Z, s\le t}\psi_\varepsilon(s)$ if $t \in \mathbb{R}$ and $(-\infty,t]\cap Z \neq \varnothing$, and $\widetilde{\psi}_\varepsilon(t) = \inf_{s\in Z}\psi_\varepsilon(s)$ otherwise. Then $\widetilde{\psi}_\varepsilon : \mathbb{R} \to \{0,1,\dots,M(\varepsilon)\}$ is a nondecreasing step function with a finite set $P_\varepsilon \subset \mathbb{R}$ of points of discontinuity. Let us show that

$$\lim_{j,k\to\infty} N(\varepsilon, f_j^\varepsilon, f_k^\varepsilon, (-\infty,t]\cap T) = \widetilde{\psi}_\varepsilon(t) \quad \text{for all} \quad t \in T\setminus P_\varepsilon. \tag{33}$$

Taking into account (32), we may assume that $t \in T\setminus(P_\varepsilon\cup Z)$. Since $t$ is a point of continuity of $\widetilde{\psi}_\varepsilon$ and $\widetilde{\psi}_\varepsilon$ is discrete valued, there exists a $\delta = \delta(\varepsilon,t) > 0$ such that $\widetilde{\psi}_\varepsilon(s) = \widetilde{\psi}_\varepsilon(t)$ for all $t-\delta \le s \le t+\delta$. Moreover, since $t \notin L_T$ and $T$ is a subset of the closure of $Z$, we have $\varnothing \neq (t-\delta,t)\cap T \subset (t-\delta,t)\cap(\text{closure of } Z)$, and so, one can find an $s_1 = s_1(\varepsilon,t)$ in $(t-\delta,t)\cap Z$. In a similar manner $t\notin R_T$ implies the existence of a point $s_2 = s_2(\varepsilon,t) \in Z$ such that $t < s_2 < t+\delta$. It follows from (32) that

$$\widetilde{\psi}_\varepsilon(t) = \widetilde{\psi}_\varepsilon(s_1) = \psi_\varepsilon(s_1) = \lim_{j,k\to\infty} N(\varepsilon, f_j^\varepsilon, f_k^\varepsilon, (-\infty,s_1]\cap T),$$

$$\widetilde{\psi}_\varepsilon(t) = \widetilde{\psi}_\varepsilon(s_2) = \psi_\varepsilon(s_2) = \lim_{j,k\to\infty} N(\varepsilon, f_j^\varepsilon, f_k^\varepsilon, (-\infty,s_2]\cap T),$$

and, by virtue of inequalities $s_1 < t < s_2$ and Lemma 3.1(c),

$$N(\varepsilon, f_j^\varepsilon, f_k^\varepsilon, (-\infty,s_1]\cap T) \le N(\varepsilon, f_j^\varepsilon, f_k^\varepsilon, (-\infty,t]\cap T) \le N(\varepsilon, f_j^\varepsilon, f_k^\varepsilon, (-\infty,s_2]\cap T)$$

for $j\neq k$, whence (33) follows.

In order to finish the proof of (29), we note that $T = (T\setminus P_\varepsilon)\cup(T\cap P_\varepsilon)$, where $T\cap P_\varepsilon$ is finite, and apply the arguments proving (32) with $Z$ replaced by $T\cap P_\varepsilon$ and $\{f_j\}$—by $\{f_j^\varepsilon\}$: there exist a subsequence of $\{f_j^\varepsilon\}$, again denoted by $\{f_j^\varepsilon\}$, and a nondecreasing function $\chi_\varepsilon : T\cap P_\varepsilon \to \{0,1,\dots,M(\varepsilon)\}$ such that the double limit on the left in (33) is equal to $\chi_\varepsilon(t)$ for all $t\in T\cap P_\varepsilon$. Defining $N_\varepsilon : T \to \{0,1,\dots,M(\varepsilon)\}$ by $N_\varepsilon(t) = \widetilde{\psi}_\varepsilon(t)$ if $t\in T\setminus P_\varepsilon$ and $N_\varepsilon(t) = \chi_\varepsilon(t)$ if $t \in T\cap P_\varepsilon$, we arrive at (29) where, in view of Lemma 3.1(c), the function $N_\varepsilon$ is nondecreasing on $T$.

2. Making use of the diagonal process here we prove that, given a decreasing sequence of positive numbers $\{\alpha_m\}_{m=1}^\infty$ tending to zero, there is a subsequence of $\{f_j\}$, again denoted by $\{f_j\}$, and for each $m \in \mathbb{N}$ there exists a nondecreasing function $K_m : T \to \{0,1,\dots,M(\alpha_m)\}$ such that

$$\lim_{j,k\to\infty} N(\alpha_m, f_j, f_k, (-\infty,t]\cap T) = K_m(t) \quad \text{for all } m\in\mathbb{N} \text{ and } t\in T. \tag{34}$$

By step 1 with $\varepsilon = \alpha_1$, there are a subsequence $\{f_j^{(1)}\}_{j=1}^\infty$ of the original sequence $\{f_j\}$ and a nondecreasing function $K_1 = N_{\alpha_1} : T \to \{0,1,\dots,M(\alpha_1)\}$ such that

$$\lim_{j,k\to\infty} N(\alpha_1, f_j^{(1)}, f_k^{(1)}, (-\infty,t]\cap T) = K_1(t) \quad \text{for all} \quad t\in T.$$

Inductively, if $m \ge 2$ and a subsequence $\{f_j^{(m-1)}\}_{j=1}^\infty$ of $\{f_j\}$ is already chosen, by virtue of step 1 with $\varepsilon = \alpha_m$ applied to the sequence $\{f_j^{(m-1)}\}_{j=1}^\infty$ (instead of

$\{f_j\}$ there), we find a subsequence $\{f_j^{(m)}\}_{j=1}^\infty$ of $\{f_j^{(m-1)}\}_{j=1}^\infty$ and a nondecreasing function $K_m : T \to \{0, 1, \ldots, M(\alpha_m)\}$ such that

$$\lim_{j,k\to\infty} N(\alpha_m, f_j^{(m)}, f_k^{(m)}, (-\infty, t] \cap T) = K_m(t) \quad \text{for all} \quad t \in T.$$

It follows that the diagonal subsequence $\{f_j^{(j)}\}_{j=1}^\infty$ of $\{f_j\}$, again denoted by $\{f_j\}$, satisfies property (34).

3. In this step we complete the proof of Theorem 2.2. Let $Q$ denote an at most countable dense subset of $T$ and $Q_m \subset T$ the set of all points of discontinuity of the nondecreasing function $K_m$, $m \in \mathbb{N}$. Then the set $S = Q \cup \bigcup_{m=1}^\infty Q_m$ is an at most countable dense subset of $T$ having the property: each function $K_m$ is continuous on $T \setminus S$. Since the sequence $\{f_j(t)\}$ is precompact in $X$ for all $t \in T$ and the set $S \subset T$ is at most countable, we may assume (applying the standard diagonal process and passing to a subsequence of $\{f_j\}$ from step 2 if necessary) that, for all $s \in S$, $f_j(s)$ converges in $X$ as $j \to \infty$ to a point of $X$ denoted by $f(s)$.

It remains to prove that the sequence $\{f_j\}$ is convergent at points of $T \setminus S$. Let $t \in T \setminus S$, and $\varepsilon > 0$ be arbitrarily fixed. Choose and fix an $m = m(\varepsilon) \in \mathbb{N}$ such that $\alpha_m \leq \varepsilon$. The definition of $S$ implies that $t$ is a limit point for $T$ and a point of continuity of $K_m$, and so, by the density of $S$ in $T$ there exists an $s = s(m,t) \in S$ such that $|K_m(t) - K_m(s)| < 1$ or $K_m(t) = K_m(s)$. Property (34) yields the existence of two positive integers $j_1 = j_1(m,t)$ and $j_2 = j_2(m,s)$ such that if $j, k \geq \max\{j_1, j_2\}$ and $j \neq k$, then

$$N(\alpha_m, f_j, f_k, (-\infty, t] \cap T) = K_m(t) \quad \text{and} \quad N(\alpha_m, f_j, f_k, (-\infty, s] \cap T) = K_m(s).$$

Assuming that $s < t$ (the arguments for $t < s$ are similar) and applying Lemma 3.1(e), for all $j, k \geq \max\{j_1, j_2\}$, $j \neq k$, we get:

$$\begin{aligned} N(\alpha_m, f_j, f_k, [s,t] \cap T) &\leq N(\alpha_m, f_j, f_k, (-\infty, t] \cap T) \\ &\qquad - N(\alpha_m, f_j, f_k, (-\infty, s] \cap T) \\ &= K_m(t) - K_m(s) = 0. \end{aligned}$$

Therefore, $N(\alpha_m, f_j, f_k, [s,t] \cap T) = 0$, which, by definition (8), yields

$$d(f_j(t) + f_k(s), f_k(t) + f_j(s)) \leq \alpha_m \leq \varepsilon \quad \text{for all} \quad j, k \geq \max\{j_1, j_2\}.$$

Since $\{f_j(s)\}$ is convergent in $X$, it is a Cauchy sequence, and so, there is a positive integer $j_3 = j_3(\varepsilon, s)$ such that $d(f_k(s), f_j(s)) \leq \varepsilon$ for all $j, k \geq j_3$. Then the integer $j_4 = \max\{j_1, j_2, j_3\}$ depends only on $\varepsilon$ (and $t$), and by virtue of (26) we have:

$$d(f_j(t), f_k(t)) \leq d(f_j(t)+f_k(s), f_k(t)+f_j(s)) + d(f_k(s), f_j(s)) \leq \varepsilon + \varepsilon = 2\varepsilon$$

for all $j, k \geq j_4$. Thus, $\{f_j(t)\}$ is a Cauchy sequence in $X$ and, since it is precompact in $X$ by assumption, it is convergent in $X$ to a point denoted by $f(t)$. Noting that $T = S \cup (T \setminus S)$ we find that the sequence $\{f_j\}$ converges pointwise on $T$ to $f$, which completes the proof of Theorem 2.2. □

A number of remarks concerning Theorem 2.2 are in order.

**Remark 5.1.** If $\{f_j\} \subset X^T$, $f \in X^T$ and $f_j \rightrightarrows f$ on $T$, then

$$\lim_{j,k\to\infty} N(\varepsilon, f_j, f_k, T) = 0 \quad \text{for all} \quad \varepsilon > 0. \tag{35}$$

In fact, given $\varepsilon > 0$, there exists a $j_0 = j_0(\varepsilon) \in \mathbb{N}$ such that $d(f_j(t), f_k(t)) \le \varepsilon/2$ for all $j, k \ge j_0(\varepsilon)$ and $t \in T$. Then it follows from (27) that

$$d(f_j(t) + f_k(s), f_k(t) + f_j(s)) \le d(f_j(t), f_k(t)) + d(f_k(s), f_j(s)) \le \varepsilon$$

for all $j, k \ge j_0(\varepsilon)$ and $s, t \in T$, and so, $N(\varepsilon, f_j, f_k, T) = 0$ for all $j, k \ge j_0(\varepsilon)$.

However, Example 5 constructed in [16, Section 4] shows that condition (10) can be fulfilled in its full generality for the pointwise convergence of $f_j$ to $f$.

**Remark 5.2.** Applying (35) and the arguments from the proof of Theorem 4.1(d), we get the following: if $\varnothing \ne T \subset \mathbb{R}$ is a measurable set with finite Lebesgue measure and $\{f_j\} \subset X^T$ is a sequence of measurable functions which converges pointwise or a.e. on $T$, then for each $\eta > 0$ there exists a measurable set $E_\eta \subset T$ of measure $\le \eta$ such that

$$\lim_{j,k\to\infty} N(\varepsilon, f_j, f_k, T \setminus E_\eta) = 0 \quad \text{for all} \quad \varepsilon > 0.$$

**Remark 5.3.** Applying Theorem 2.2 and the diagonal process we obtain a corollary of this theorem similar to Corollary 4.1; we omit the formulation and the proof. Also, in the same manner, from Theorem 2.2, one obtains a result for the a.e. convergence similar to Theorem 4.2.

**Remark 5.4.** The *joint modulus of variation* $\{\nu(n, f, g, T)\}_{n=1}^\infty$ (see [16, equality (5)]) and the *joint oscillation* $\operatorname{osc}(f, g, T)$ of $f, g \in X^T$ are introduced as follows (cf. (2)):

$$\nu(n, f, g, T) = \sup\Big\{\sum_{i=1}^n |f, g(I_i)| : \{I_i\}_1^n \prec T\Big\}, \qquad n \in \mathbb{N},$$

$$\operatorname{osc}(f, g, T) = \sup_{I \prec T} |f, g(I)| = \sup_{s,t\in T} d(f(t) + g(s), g(t) + f(s)).$$

Since the double limit superior (9) mimics a property of the ordinary limit superior (i.e., $\limsup_{j\to\infty} \alpha_j = \lim_{k\to\infty} \sup_{j\ge k} \alpha_j$), we have the following counterpart of Lemma 3.3(c): if $\{f_j\} \subset X^T$, then $\limsup_{j,k\to\infty} \nu(n, f_j, f_k, T) = o(n)$ iff $\limsup_{j,k\to\infty} N(\varepsilon, f_j, f_k, T) < \infty$ for all $\varepsilon > 0$ and $\limsup_{j,k\to\infty} \operatorname{osc}(f_j, f_k, T) < \infty$. In this way Theorem 2.2 implies Theorem 1 from Chistyakov and Maniscalco [16], where instead of (10) condition $\limsup_{j,k\to\infty} \nu(n, f_j, f_k, T) = o(n)$ was assumed in order to obtain the conclusion of Theorem 2.2 above.

**Remark 5.5.** Given $f, g \in X^T$, the *joint $\varphi$-variation* $V_\varphi(f, g, T)$ in Gniłka's sense and the *joint $\Phi$-variation* $V_\Phi(f, g, T)$ in Schramm's sense are introduced as in Section 3 by replacing $f$ by the pair $f, g$. Then the modified estimates (15)–(18) and

Theorem 2.2 imply the following assertion: the conclusion of Theorem 2.2 holds if condition (10) is replaced by $\limsup_{j,k\to\infty} V_*(f_j, f_k, T) < \infty$ with $* = \varphi$ or $\Phi$.

**Remark 5.6.** The sequence $f_j(t) = ((-1)^j + (1/j))\mathcal{D}(t)$, $t \in [0,1]$, $j \in \mathbb{N}$, satisfies the conditions of [21, Theorem 2.1], but not of Theorem 2.2 above. This and the examples from Maniscalco [28] show that if $X = \mathbb{R}$, then our Theorem 2.2 on the one hand and Theorem 1.2 from Schrader [33] and Theorem 2.1 from Di Piazza and Maniscalco [21] on the other hand are independent.

## 6. Functions with Values in a Reflexive Separable Banach Space

The aim of this section is to establish several variants of Theorem 2.1 for the weak pointwise convergence of sequences of functions whose values lie in a reflexive separable Banach space, taking into account certain specific features of this case.

Let $(X, \|\cdot\|)$ be a normed linear space over the field $\mathbb{K} = \mathbb{R}$ or $\mathbb{C}$ and $X^*$ be its dual, i. e., the space $L(X;\mathbb{K})$ of all continuous linear functionals on $X$. The space $X^*$ is a Banach space with respect to the norm $\|x^*\| = \sup\{|x^*(x)| : x \in X$ and $\|x\| \le 1\}$ for all $x^* \in X^*$. The bilinear functional $\langle\cdot,\cdot\rangle : X \times X^* \to \mathbb{K}$ defined by $\langle x, x^*\rangle = x^*(x)$ for all $x \in X$ and $x^* \in X^*$ gives rise to the natural duality between $X$ and $X^*$. Recall that a sequence $\{x_j\} \subset X$ is said to be *weakly convergent* in $X$ to $x \in X$, which is written as $x_j \xrightarrow{w} x$ in $X$, provided $\langle x_j, x^*\rangle \to \langle x, x^*\rangle$ in $\mathbb{K}$ as $j \to \infty$ for all $x^* \in X^*$; if this is the case, we have $\|x\| \le \liminf_{j\to\infty} \|x_j\|$.

The definition of the quantity $N(\varepsilon, f, T)$ for $f \in X^T$ is the same as in Section 2 with respect to the induced metric $d$ on $X$, i. e., $d(x,y) = \|x-y\|$ for $x$, $y \in X$. The following theorem is a generalization of Theorem 7 from [13].

**Theorem 6.1.** *Let $\varnothing \ne T \subset \mathbb{R}$ and $(X, \|\cdot\|)$ be a reflexive separable Banach space with separable dual $X^*$. Suppose the sequence of functions $\{f_j\} \subset X^T$ is such that $\sup_{j\in\mathbb{N}} \|f_j(t)\| < \infty$ for all $t \in T$, and condition (6) holds. Then $\{f_j\}$ contains a subsequence, again denoted by $\{f_j\}$, and there is a function $f \in X^T$ satisfying $N(\varepsilon, f, T) \le N(\varepsilon)$ for all $\varepsilon > 0$ such that $f_j(t) \xrightarrow{w} f(t)$ in $X$ for all $t \in T$.*

**Proof.** We set $C(t) = \sup_{j\in\mathbb{N}} \|f_j(t)\| < \infty$, $t \in T$, and note that, given $j \in \mathbb{N}$ and $x^* \in X^*$, we have:

$$|\langle f_j(t), x^*\rangle| \le \|f_j(t)\| \cdot \|x^*\| \le C(t)\|x^*\|, \qquad t \in T, \tag{36}$$

and

$$N(\varepsilon, \langle f_j(\cdot), x^*\rangle, T) \begin{cases} \le N(\varepsilon/\|x^*\|, f_j, T) & \text{if } x^* \ne 0, \\ = 0 & \text{if } x^* = 0. \end{cases}$$

The latter condition and (6) imply that

$$N_{x^*}(\varepsilon) \equiv \limsup_{j\to\infty} N(\varepsilon, \langle f_j(\cdot), x^*\rangle, T) \begin{cases} \le N(\varepsilon/\|x^*\|) < \infty & \text{if } x^* \ne 0, \\ = 0 & \text{if } x^* = 0. \end{cases} \tag{37}$$

Let $\{x_k^*\}_{k=1}^\infty$ be a countable dense subset of $X^*$. Setting $x^* = x_1^*$ in (36) and (37) and applying Theorem 2.1, we get a subsequence $\{f_j^{(1)}\}_{j=1}^\infty$ and a function $y_{x_1^*} \in \mathbb{K}^T$ depending on $x_1^*$ such that $N(\varepsilon, y_{x_1^*}, T) \leq N_{x_1^*}(\varepsilon)$ for all $\varepsilon > 0$ and $\langle f_j^{(1)}(t), x_1^* \rangle \to y_{x_1^*}(t)$ in $\mathbb{K}$ for all $t \in T$. If $k \geq 2$ and a subsequence $\{f_j^{(k-1)}\}_{j=1}^\infty$ of $\{f_j\}$ is already chosen, by (36) and (37) with $x^* = x_k^*$ and Theorem 2.1, we have:

$$\sup_{j \in \mathbb{N}} |\langle f_j^{(k-1)}(t), x_k^* \rangle| \leq C(t)\|x_k^*\| \quad \text{for all} \quad t \in T, \quad \text{and}$$

$$\limsup_{j \to \infty} N(\varepsilon, \langle f_j^{(k-1)}(\cdot), x_k^* \rangle, T) \leq N_{x_k^*}(\varepsilon) < \infty \quad \text{for all} \quad \varepsilon > 0,$$

and so, Theorem 2.1 implies the existence of a subsequence $\{f_j^{(k)}\}_{j=1}^\infty$ of $\{f_j^{(k-1)}\}_{j=1}^\infty$ and a function $y_{x_k^*} \in \mathbb{K}^T$ satisfying $N(\varepsilon, y_{x_k^*}, T) \leq N_{x_k^*}(\varepsilon)$ for all $\varepsilon > 0$ such that $\langle f_j^{(k)}(t), x_k^* \rangle \to y_{x_k^*}(t)$ in $\mathbb{K}$ for all $t \in T$. Then the diagonal subsequence $\{f_j^{(j)}\}_{j=1}^\infty$ of $\{f_j\}$, again denoted by $\{f_j\}$, satisfies the condition:

$$\langle f_j(t), x_k^* \rangle \to y_{x_k^*}(t) \quad \text{in } \mathbb{K} \text{ as } j \to \infty \text{ for all } t \in T \text{ and } k \in \mathbb{N}.$$

Since the sequence $\{x_k^*\}_{k=1}^\infty$ is dense in $X^*$, by the standard arguments (cf. step (3) in the proof of [13, Theorem 7]) one shows that for each $x^* \in X^*$ there exists a function $y_{x^*} \in \mathbb{K}^T$ satisfying $N(\varepsilon, y_{x^*}, T) \leq N_{x^*}(\varepsilon)$ for all $\varepsilon > 0$ such that

$$\langle f_j(t), x^* \rangle \to y_{x^*}(t) \quad \text{in } \mathbb{K} \text{ as } j \to \infty \text{ for all } t \in T \text{ and } x^* \in X^*. \tag{38}$$

By the reflexivity of $X$, the uniform boundedness principle and the standard procedure (e.g., such as in step (4) of the proof of [13, Theorem 7]), defining $f : T \to L(X^*; \mathbb{K}) = X^{**} = X$ by $\langle f(t), x^* \rangle = y_{x^*}(t)$ for all $t \in T$ and $x^* \in X^*$, we find from (38) that $f \in X^T$, $\|f(t)\| \leq \liminf_{j\to\infty} \|f_j(t)\|$ for all $t \in T$ and $\langle f_j(t), x^* \rangle \to \langle f(t), x^* \rangle$ in $\mathbb{K}$ as $j \to \infty$ for all $x^* \in X^*$ and $t \in T$, i.e., $f_j(t) \xrightarrow{w} f(t)$ in $X$ for all $t \in T$.

It remains to prove that $N(\varepsilon, f, T) \leq N(\varepsilon)$ for all $\varepsilon > 0$. Assuming that $N(\varepsilon, f, T) > 0$, let $n \in \mathbb{N}$ be such that there exists $\{I_i\}_1^n \prec T$ with $I_i = \{s_i, t_i\}$ and $|f(I_i)| = \|f(s_i) - f(t_i)\| > \varepsilon$ for all $i = 1, \ldots, n$. Since $f_j(s_i) - f_j(t_i) \xrightarrow{w} f(s_i) - f(t_i)$ in $X$ as $j \to \infty$, we have

$$\varepsilon < \|f(s_i) - f(t_i)\| \leq \liminf_{j\to\infty} \|f_j(s_i) - f_j(t_i)\|,$$

and so, there exists $j_0(\varepsilon, n) \in \mathbb{N}$ such that $\inf_{j \geq j_0(\varepsilon,n)} \|f_j(s_i) - f_j(t_i)\| > \varepsilon$ for all $i = 1, \ldots, n$. By virtue of (4), it follows that $n \leq N(\varepsilon, f_j, T)$ for all $j \geq j_0(\varepsilon, n)$. This gives

$$n \leq \inf_{j \geq j_0(\varepsilon,n)} N(\varepsilon, f_j, T) \leq \liminf_{j\to\infty} N(\varepsilon, f_j, T)$$

for all $n$ with the properties as above. Thus,

$$N(\varepsilon, f, T) \leq \liminf_{j\to\infty} N(\varepsilon, f_j, T) \leq \limsup_{j\to\infty} N(\varepsilon, f_j, T) \leq N(\varepsilon),$$

which was to be proved. □

It is to be noted that, by virtue of estimates (15)–(18), condition (6) in Theorem 6.1 can be replaced by a *stronger* condition: the value $V_* \equiv \sup_{j\in\mathbb{N}} V_*(f_j, T)$ is finite, where $* = \varphi$ or $* = \Phi$; in this case the weak pointwise limit function $f \in \mathrm{BV}_*(T; X)$ from Theorem 6.1 is such that $V_*(f, T) \le V_*$.

The next theorem can be proved by following the proof of Theorem 4.2 and applying Theorem 6.1 in place of Theorem 2.1.

**Theorem 6.2.** *Let $T$ and $X$ be as in Theorem 6.1 and the sequence $\{f_j\} \subset X^T$ satisfy the conditions: (i) $\sup_{j\in\mathbb{N}} \|f_j(t)\| < \infty$ for almost all $t \in T$, and (ii) for each $k \in \mathbb{N}$ there exists a measurable set $E_k \subset T$ of measure $\le 1/k$ such that condition (25) holds. Then there exist a subsequence of $\{f_j\}$, again denoted by $\{f_j\}$, and a function $f \in X^T$, having the property that for each $k \in \mathbb{N}$ there is a measurable set $E'_k \subset T$ of measure $\le 1/k$ for which $N(\varepsilon, f, T \setminus E'_k) < \infty$ for all $\varepsilon > 0$, such that $f_j(t) \xrightarrow{w} f(t)$ in $X$ for almost all $t \in T$.*

**Remark 6.1.** By virtue of Theorem 2.2, if condition (6) in Theorem 6.1 is replaced by condition (10), then $\{f_j\}$ contains a subsequence, again denoted by $\{f_j\}$, such that $f_j(t) \xrightarrow{w} f(t)$ in $X$ for all $t \in T$ and some function $f \in X^T$. Analogously, Theorem 6.2 can be adapted for the case of the joint (25)-condition in (ii). We omit the details.

## Acknowledgments

The initial version of this paper has been written when the first author was visiting Dipartimento di Matematica ed Applicazioni della Università degli Studi di Palermo, Palermo, Italy, September 23–October 6, 2007. It is a pleasure to thank Benedetto Bongiorno, Luisa Di Piazza and Department of Mathematics for their support. Special thanks go to Gianni Morando for his kind hospitality.

## References

[1] S. A. Belov, V. V. Chistyakov, *A selection principle for mappings of bounded variation*, J. Math. Anal. Appl. **249**, no. 2 (2000), 351–366.

[2] S. A. Belov, V. V. Chistyakov, *Regular selections of multifunctions of bounded variation*, J. Math. Sci. (New York) **110**, no. 2 (2002), 2452–2454.

[3] Z. A. Chanturiya, *The modulus of variation of a function and its application in the theory of Fourier series*, Dokl. Akad. Nauk SSSR **214**, no. 1 (1974), 63–66 (in Russian). English translation: Soviet Math. Dokl. **15**, no. 1 (1974), 67–71.

[4] Z. A. Chanturiya, *Absolute convergence of Fourier series*, Mat. Zametki **18**, no. 2 (1975), 185–192 (in Russian). English translation: Math. Notes **18**, no. 1 (1975), 695–700.

[5] V. V. Chistyakov, *On mappings of bounded variation*, J. Dynam. Control Systems **3**, no. 2 (1997), 261–289.

[6] V. V. Chistyakov, *On the theory of multivalued mappings of bounded variation of one real variable*, Mat. Sb. **189**, no. 5 (1998), 153–176 (in Russian). English translation: Sbornik Math. **189**, no. 5–6 (1998), 797–819.

[7] V. V. Chistyakov, *Mappings of bounded variation with values in a metric space: generalizations*, J. Math. Sci. (New York) **100**, no. 6 (2000), 2700–2715.

[8] V. V. Chistyakov, *Generalized variation of mappings with applications to composition operators and multifunctions*, Positivity **5**, no. 4 (2001), 323–358.

[9] V. V. Chistyakov, *On multi-valued mappings of finite generalized variation*, Mat. Zametki **71**, no. 4 (2002), 611–632 (in Russian). English translation: Math. Notes **71**, no. 3–4 (2002), 556–575.

[10] V. V. Chistyakov, *Selections of bounded variation*, J. Appl. Anal. **10**, no. 1 (2004), 1–82.

[11] V. V. Chistyakov, *A new pointwise selection principle for mappings of one real variable*, Intern. School-Confer. in Analysis and Geometry dedicated to the 75th Birthday of Academician Yu. G. Reshetnyak. Novosibirsk, Russia (2004), 30–31.

[12] V. V. Chistyakov, *A selection principle for functions of a real variable*, Atti Semin. Mat. Fis. Univ. Modena Reggio Emilia **53**, no. 1 (2005), 25–43.

[13] V. V. Chistyakov, *The optimal form of selection principles for functions of a real variable*, J. Math. Anal. Appl. **310**, no. 2 (2005), 609–625.

[14] V. V. Chistyakov, *A selection principle for functions with values in a uniform space*, Dokl. Akad. Nauk **409**, no. 5 (2006), 591–593 (in Russian). English translation: Dokl. Math. **74**, no. 1 (2006), 559–561.

[15] V. V. Chistyakov, *A pointwise selection principle for functions of one variable with values in a uniform space*, Mat. Trudy **9**, no. 1 (2006), 176–204 (in Russian). English translation: Siberian Adv. Math. **16**, no. 3 (2006), 15–41.

[16] V. V. Chistyakov, C. Maniscalco, *A pointwise selection principle for metric semigroup valued functions*, J. Math. Anal. Appl. **341**, no. 1 (2008), 613–625.

[17] V. V. Chistyakov, A. Nowak, *Regular Carathéodory-type selectors under no convexity assumptions*, J. Funct. Anal. **225**, no. 2 (2005), 247–262.

[18] V. V. Chistyakov, D. Repovš, *Selections of bounded variation under the excess restrictions*, J. Math. Anal. Appl. **331**, no. 2 (2007), 873–885.

[19] V. V. Chistyakov, A. Rychlewicz, *On the extension and generation of set-valued mappings of bounded variation*, Studia Math. **153**, no. 3 (2002), 235–247.

[20] V. V. Chistyakov, Yu. V. Tretyachenko, *A selection principle for pointwise bounded sequences of functions*, Math. Notes (2008) (to appear).

[21] L. Di Piazza, C. Maniscalco, *Selection theorems, based on generalized variation and oscillation*, Rend. Circ. Mat. Palermo, Ser. II, **35** (1986), 386–396.

[22] R. M. Dudley, R. Norvaiša, *Differentiability of Six Operators on Nonsmooth Functions and p-Variation* (with the collaboration of Jinghua Qian), Lecture Notes in Math., Vol. 1703, Springer-Verlag, Berlin, 1999.

[23] S. Fuchino, Sz. Plewik, *On a theorem of E. Helly*, Proc. Amer. Math. Soc. **127**, no. 2 (1999), 491–497.

[24] S. Gniłka, *On the generalized Helly's theorem*, Funct. Approx. Comment. Math. **4** (1976), 109–112.

[25] C. Goffman, G. Moran, D. Waterman, *The structure of regulated functions*, Proc. Amer. Math. Soc. **57**, no. 1 (1976), 61–65.

[26] C. Goffman, T. Nishiura, D. Waterman, *Homeomorphisms in Analysis*, Math. Surveys Monogr., Vol. 54, Amer. Math. Soc., Providence, RI, 1997.

[27] E. Helly, *Über lineare Funktionaloperationen*, Sitzungsber. Naturwiss. Kl. Kaiserlichen Akad. Wiss. Wien 121 (1912), 265–297.

[28] C. Maniscalco, *A comparison of three recent selection theorems*, Math. Bohem. **132**, no. 2 (2007), 177–183.

[29] J. Musielak, W. Orlicz, *On generalized variations* (I), Studia Math. **18** (1959), 11–41.

[30] S. Perlman, *Functions of generalized variation*, Fund. Math. **105** (1980), 199–211.
[31] F. Ramsey, *On a problem of formal logic*, Proc. London Math. Soc. (2) **30** (1930), 264–286.
[32] S. Saks, *Theory of the Integral*, 2nd revised edition, Stechert, New York 1937.
[33] K. Schrader, *A generalization of the Helly selection theorem*, Bull. Amer. Math. Soc. **78**, no. 3 (1972), 415–419.
[34] M. Schramm, *Functions of $\Phi$-bounded variation and Riemann-Stieltjes integration*, Trans. Amer. Math. Soc. **287**, no. 1 (1985), 49–63.
[35] R. Taberski, *On the power variations and pseudovariations of positive integer orders*, Demonstratio Math. **19**, no. 4 (1986), 881–893.
[36] D. Waterman, *On convergence of Fourier series of functions of generalized bounded variation*, Studia Math. **44**, no. 1 (1972), 107–117.
[37] D. Waterman, *On $\Lambda$-bounded variation*, Studia Math. **57**, no. 1 (1976), 33–45.
[38] N. Wiener, *The quadratic variation of a function and its Fourier coefficients*, J. Math. Phys. **3** (1924), 73–94. Reprinted in: N. Wiener, *Collected Works*. Vol. II. MIT Press, Cambridge, 1979, 36–58.
[39] L. C. Young, *General inequalities for Stieltjes integrals and the convergence of Fourier series*, Math. Ann. **115** (1938), 581–612.

# LOCAL $L^p$ INEQUALITIES FOR GEGENBAUER POLYNOMIALS

LAURA DE CARLI

*Department of Mathematics, Florida International University,*
*Miami, FL 33199*
*E-mail: decarlil@fiu.edu*

In this paper we prove new $L^p$ estimates for Gegenbauer polynomials $P_n^{(s)}(x)$. We let $d\mu_s(x) = (1-x^2)^{s-\frac{1}{2}}dx$ be the measure in $(-1,1)$ which makes the polynomials $P_n^{(s)}(x)$ orthogonal, and we compare the $L^p(d\mu_s)$ norm of $P_n^{(s)}(x)$ with that of $x^n$. We also prove new $L^p(d\mu_s)$ estimates of the restriction of these polynomials to the intervals $[0, z_n]$ and $[z_n, 1]$ where $z_n$ denotes the largest zero of $P_n^{(s)}(x)$

*Keywords*: Gegenbauer polynomials, $L^p$ estimates

## 1. Introduction

In this paper we will prove new $L^p$ estimates for Gegenbauer, (or ultraspherical), polynomials.

The Gegenbauer polynomial of order $s$ and degree $n$, $P_n^{(s)}(x)$, can be defined, for example, as the coefficients of $\omega^n$ in the expansion of the generating function $(1 - 2x\omega + w^2)^{-s} = \sum_{n=0}^{\infty} \omega^n P_n^{(s)} n(x)$. Gegenbauer polynomials are orthogonal in $L^2(-1,\ 1)$ with the measure $d\mu_s(x) = (1 - x^2)^{s-\frac{1}{2}}dx$. Other properties of these polynomials are listed in the next Section.

In this paper we aim to estimate the $L^p(d\mu_s)$ norm of Gegenbauer polynomials and the $L^p(d\mu_s)$ norm of their restrictions to certain intervals of $[-1,1]$ in terms of the $L^p(d\mu_s)$ norm of $x^n$.

This choice is motivated by the fact that $\lim_{s\to\infty} \widetilde{P}_n^{(s)}(x) = x^n$. This is easy to prove using e.g. the explicit representation (10). In [D] the sharp inequality

$$|P_n^{(s)}(x)| \le P_n^{(s)}(1) \left( |x|^n + \frac{n-1}{2s+1}(1 - |x|^n) \right) \tag{1}$$

has been proved for Gegenbauer polynomials of order $s \ge n\dfrac{1+\sqrt{5}}{4}$.

A pointwise comparison between $\widetilde{P}_n^{(s)}(x) = \dfrac{P_n^{(s)}(x)}{P_n^{(s)}(1)}$ and $x^n$ is meaningful only when $s$ is much larger that $n$.

Gegenbauer polynomials of large degree behave like Bessel functions, in the sense that

$$\lim_{n\to\infty}\frac{P_n^{(s)}\left(\cos\frac{z}{n}\right)}{P_n^{(s)}(1)}=\Gamma\left(s+\frac{1}{2}\right)\left(\frac{z}{2}\right)^{-s+\frac{1}{2}}J_{s-\frac{1}{2}}(z). \tag{2}$$

(2) easily follows from a well known Mehler-Heine type asymptotic formula for general Jacobi polynomials, (see [Sz], pg. 167).

However, $\widetilde{P}_n^{(s)}(x)$ and $x^n$ have the same $L^\infty$ norm for every $s>0$ and every $n\geq 0$. Indeed,

$$\sup_{x\in[-1,1]}\left|\widetilde{P}_n^{(s)}(x)\right|=\sup_{x\in[-1,1]}|x^n|=1$$

because $|P_n^{(s)}(x)|\leq P_n^{(s)}(1)$, (see the next Section).

Also the ratio between the $L^2(d\mu_s)$ norm of $\widetilde{P}_n^{(s)}(x)$ and the $L^2(d\mu_s)$ norm of $x^n$ can be estimated for every $n$ and $s$.

We prove the following

**Proposition 1.1.** *The function* $N_2(n,s)=\dfrac{||\widetilde{P}_n^{(s)}||_{L^2(d\mu_s)}}{||x^n||_{L^2(d\mu_s)}}$ *is decreasing with* $s$, *and*

$$2^{-\frac{n}{2}}\left(\frac{\sqrt{\pi}\Gamma(n+1)}{\Gamma\left(n+\frac{1}{2}\right)}\right)^{\frac{1}{2}}=\lim_{s\to\infty}N_2(n,s)<N_2(n,s)\leq\lim_{s\to 0}N_2(n,s)=\left(\frac{\sqrt{\pi}\Gamma(n+1)}{2\Gamma\left(n+\frac{1}{2}\right)}\right)^{\frac{1}{2}}. \tag{3}$$

*Thus,*

$$2^{-\frac{n}{2}}\pi^{\frac{1}{4}}n^{\frac{1}{4}}<N_2(n,s)<n^{\frac{1}{4}}. \tag{4}$$

It is interesting to observe that $N_2(n,\frac{1}{2})=1$. This follows from the explicit formula for $N_2(n,s)$ in Section 2. By Proposition 1.1, $N_2(n,s)=1$ if and only if $s=\frac{1}{2}$.

Proposition 1.1 shows that while it is true that $\lim_{s\to\infty}\widetilde{P}_n^{(s)}(x)=x^n$, and $\lim_{s\to\infty}||\widetilde{P}_n^{(s)}(x)||_{L^\infty(d\mu_s)}=||x^n||_{L^\infty(d\mu_s)}$, it is not true in general that $\lim_{s\to\infty}||\widetilde{P}_n^{(s)}||_{L^2(d\mu_s)}=||x^n||_{L^2(d\mu_s)}$.

These consideration suggested us to investigate the ratio of the $L^r(d\mu_s)$ norms of $\widetilde{P}_n^{(s)}(x)$ and $x^n$ for other values of $r$. We let

$$N_r(n,s)=\frac{||\widetilde{P}_n^{(s)}||_{L^r(d\mu_s)}}{||x^n||_{L^r(d\mu_s)}},\quad 1\leq r\leq\infty.$$

Our next Lemma suggests that $N_r(n,s)$ can be bounded above by a power of $N_2(n,s)$.

**Lemma 1.2.** *For every $s > 0$, $n \geq 1$, and $r \geq 2$,*

$$N_r(n,s) \leq N_2(n,s)^{\frac{2}{r}} \left(\frac{r}{2}\right)^{\frac{1}{r}(s+\frac{1}{2})}, \tag{5}$$

*and*

$$N_r(n,s) \leq n^{\frac{1}{2r}} \left(\frac{r}{2}\right)^{\frac{1}{r}(s+\frac{1}{2})}. \tag{6}$$

*When $s \to 0$ this upper bound is sharp, in the sense that the power of $n$ in (6) cannot be replaced by a smaller power.*

The proof of the Lemma is in Section 3.

Numerical evidence suggests that $N_r(n,s) \leq N_2(n,s)^{\frac{2}{r}}$ when $s \geq \frac{1}{2}$ and $1 \leq r \leq \infty$. When $0 \leq s < \frac{1}{2}$ we conjecture instead that $N_r(n,s) \geq N_2(n,s)^{\frac{2}{r}}$.

The upper bound in Lemma 1.2 can be improved if we restrict $\widetilde{P}_n^{(s)}(x)$ to the intervals $\{1 \leq |x| \leq z_n\}$ and $(-z_n, z_n)$, where $z_n$ denotes the largest positive zero of $P_n^{(s)}(x)$.

Our main result is the following.

**Theorem 1.3.** *For every $n > 2$, $s > 0$, and $r \geq 1$,*

$$\sin^{\frac{2}{r}}\left(\frac{\pi}{n+1}\right)(1-z_n^2)^{\frac{1}{r}(s+\frac{1}{2})} \leq \frac{||\widetilde{P}_n^{(s)}||_{L^r(\{1\leq|x|\leq z_n\},\ d\mu_s)}}{||x^n||_{L^r(d\mu_s)}} \leq p(n,s)^{(s+\frac{1}{2})\frac{1}{[\frac{n+1}{2}]r}}, \tag{7}$$

*where* $p(n,s) = \prod_{j=1}^{n}(1-z_j) = \dfrac{\Gamma(s)\,\Gamma(n+2\,s)}{2^n\,\Gamma(2\,s)\,\Gamma(n+s)}$ *is as in (23).*

Using Stirling's formula, it is possible to prove that $\lim_{s\to\infty} p(n,s)^{s+\frac{1}{2}} = e^{-\frac{n(n-1)}{4}}$, and thus $\lim_{s\to\infty} \dfrac{||\widetilde{P}_n^{(s)}||_{L^r(\{1\leq|x|\leq z_n\},\ d\mu_s)}}{||x^n||_{L^r(d\mu_s)}} \leq \lim_{s\to\infty} p(n,s)^{(s+\frac{1}{2})\frac{2}{nr}} = e^{-\frac{n-1}{2r}}$.

We have recalled in the next section that $z_n < \cos\left(\dfrac{\pi}{n+1}\right)\sqrt{\dfrac{(n-1)(n+2s-2)}{(n+s-2)(n+s-1)}}$, (see (20)), and so

$$\begin{aligned}
&\lim_{s\to\infty} \sin^{\frac{2}{r}}\left(\frac{\pi}{n+1}\right)(1-z_n^2)^{s+\frac{1}{2}} \\
&> \sin^{\frac{2}{r}}\left(\frac{\pi}{n+1}\right)\lim_{s\to\infty}\left(1-\frac{(n-1)(n+2s-2)\cos^2\left(\frac{\pi}{n+1}\right)}{(n+s-2)(n+s-1)}\right)^{s+\frac{1}{2}} \\
&= \sin^{\frac{2}{r}}\left(\frac{\pi}{n+1}\right)e^{-\frac{2}{r}(n-1)\cos^2\left(\frac{\pi}{n+1}\right)}.
\end{aligned}$$

From the inequalities above and (7) follows that

$$\sin^{\frac{2}{r}}\left(\frac{\pi}{n+1}\right) e^{-\frac{2}{r}(n-1)\cos^2\left(\frac{\pi}{n+1}\right)} < \lim_{s\to\infty} \frac{||\widetilde{P}_n^{(s)}||_{L^r(\{1\le|x|\le z_n\},\ d\mu_s)}}{||x^n||_{L^r(d\mu_s)}} < e^{-\frac{n-1}{2r}}. \quad (8)$$

This upper bound is not sharp; in fact we have proved in Proposition 1.1 that $\lim_{s\to\infty} N_2(n,s) = (\pi n)^{\frac{1}{4}} 2^{-\frac{n}{2}}$, while Lemma 1.3 yields $\lim_{s\to\infty} \frac{||\widetilde{P}_n^{(s)}||_{L^r(\{1\le|x|\le z_n\},\ d\mu_s)}}{||x^n||_{L^r(d\mu_s)}} \le e^{-\frac{n-1}{4}}$, and $e^{-\frac{n-1}{4}} > (\pi n)^{\frac{1}{4}} 2^{-\frac{n}{2}}$ for every $n \ge 2$. However, Theorem (1.3) is interesting because it provides an upper and lower bound for the $L^r(\{1 \le |x| \le z_n\},\ d\mu_s)$ norm of $\widetilde{P}_n^{(s)}(x)$ and is valid for every $r \ge 1$.

Since $\lim_{s\to\infty} z_n = 0$, (see the next Section), it is natural to conjecture that $N_r(n,s)$ is bounded above by a constant independent of $s$. In order to prove this conjecture we will prove that the ratio of the $L^r(d\mu_s)$ norm of $\widetilde{P}_n^{(s)}(x)$ in $(-z_n, z_n)$ and $||x^n||_{L^r(d\mu_s)}$ is a bounded function of $s$.

In the next Theorem we estimate the $L^r(d\mu_s)$ norm of $\widetilde{P}_n^{(s)}(x)$ in $(-z_n, z_n)$ through interpolation.

**Theorem 1.4.** *For every $r \ge 2$, $s > 0$ and $n \ge 2$,*

$$\frac{||\widetilde{P}_n^{(s)}||_{L^r((-z_n,\ z_n),\ d\mu_s)}}{||x^n||_{L^r(d\mu_s)}} \le N_2(n,s)^{\frac{2}{r}} \left(\frac{n(n+2s)}{2s+1}\right)^{1-\frac{2}{r}} \left(z_n^2 \frac{\frac{nr}{2}+s+1}{n-\frac{1}{2}}\right)^{n(\frac{1}{2}-\frac{1}{r})}$$

*where $z_n$ denotes the largest zero of $P_n^{(s)}$. Furthermore,*

$$\lim_{s\to\infty} \frac{||\widetilde{P}_n^{(s)}||_{L^r((-z_n,\ z_n),\ d\mu_s)}}{||x^n||_{L^r(d\mu_s)}} \le n^{1-\frac{2}{r}} 2^{n(\frac{1}{2}-\frac{1}{r})} N_2(n,s)^{\frac{2}{r}}.$$

From Theorems 1.3 and 1.4 and Proposition 1.1 we can easily prove the following

**Corollary 1.5.** *For every $n \ge 2$ and every $r \ge 2$, $\lim_{s\to\infty} N_r(n,s)$ is finite. If $r < 4$ this limit is bounded above by a constant that does not depend on $n$.*

## 2. Preliminaries

Let us briefly review the main properties of the ultraspherical polynomials. For more details we refer to [Sz]. The ultraspherical polynomials can be defined through Rodriguez' formula

$$(1-x^2)^{s-\frac{1}{2}} P_n^{(s)}(x) = \frac{(-1)^n \Gamma(s+\frac{1}{2})\Gamma(n+2s)}{\Gamma(2s)\Gamma(n+s+\frac{1}{2})\Gamma(n+1)2^n} \left(\frac{d}{dx}\right)^n (1-x^2)^{n+s-\frac{1}{2}}. \quad (9)$$

When $s > 0$ we have the following explicit expression

$$P_n^{(s)}(x) = \sum_{m=0}^{[\frac{n}{2}]} (-1)^m \frac{\Gamma(n-m+s)}{\Gamma(s)\Gamma(m+1)\Gamma(n-2m+1)} (2x)^{n-2m}. \quad (10)$$

Note that $P_n^{(s)}(x)$ is either even or odd.

Ultraspherical polynomials or order $s = 0$ are related to the Tchebicheff polynomials $T_n(x) = \cos(n \cos^{-1}(x))$ by the following limit relation.

$$\lim_{s \to 0} s^{-1} P_n^{(s)}(x) = \frac{2}{n} T_n(x). \tag{11}$$

The $L^2$ norm of $P_n^{(s)}(x)$ with respect to the measure $d\mu_s(x) = (1 - x^2)^{s-\frac{1}{2}} dx$ in $(-1, 1)$ can be explicitly computed. It is

$$||P_n^{(s)}||^2_{L^2(d\mu_s)} = \int_{-1}^{1} |P_n^{(s)}(x)|^2 (1 - x^2)^{s-\frac{1}{2}} dx = \frac{\pi 2^{1-2s} \Gamma(n + 2s)}{(n + s)(\Gamma(s))^2 \Gamma(n + 1)}. \tag{12}$$

When $s > 0$ the maximum of of $P_n^{(s)}(x)$ in $[-1, 1]$ can be explicitly computed. We have:

$$\sup_{-1 \leq x \leq 1} |P_n^{(s)}(x)| = P_n^{(s)}(1) = \frac{\Gamma(n + 2s)}{\Gamma(n + 1)\Gamma(2s)}, \quad s > 0. \tag{13}$$

$N_2(n, s)$ can be explicitly computed as well. Indeed, the $L^r(d\mu_s)$ norm of $x^n$ is

$$||x^n||_{L^r(d\mu_s)} = \left( \int_{-1}^{1} x^{nr} (1 - x^2)^{s-\frac{1}{2}} dx \right)^{\frac{1}{r}}$$

$$= B^{\frac{1}{r}} \left( \frac{1}{2}(nr + 1),\ s + \frac{1}{2} \right) = \left( \frac{\Gamma\left(\frac{1}{2}(nr + 1)\right) \Gamma\left(s + \frac{1}{2}\right)}{\Gamma\left(\frac{nr}{2} + s + 1\right)} \right)^{\frac{1}{r}}, \tag{14}$$

where $B(a, b)$ is the standard Beta function. Thus,

$$N_2(n, s) = \left( \frac{2^{-n} \sqrt{\pi} \Gamma(n + 1) \Gamma\left(s + \frac{1}{2}\right) \Gamma(n + s)}{\Gamma\left(n + \frac{1}{2}\right) \Gamma\left(\frac{n}{2} + s\right) \Gamma\left(\frac{n}{2} + s + \frac{1}{2}\right)} \right)^{\frac{1}{2}}. \tag{15}$$

This expression has been simplified with the aid of the well known duplication formula for the Gamma function $\dfrac{\Gamma(2x)}{\Gamma(x)\Gamma\left(x + \frac{1}{2}\right)} = \dfrac{2^{2x-1}}{\sqrt{\pi}}$. It is interesting to note that $N_2(n, \frac{1}{2}) \equiv 1$.

The derivatives of ultraspherical polynomials are constant multiples of ultraspherical polynomials. From (10) easily follows that

$$\frac{d}{dx} P_n^{(s)}(x) = 2s P_{n-1}^{(s+1)}(x), \tag{16}$$

and if we let $\widetilde{P}_n^{(s)}(x) = \frac{P_n^{(s)}(x)}{P_n^{(s)}(1)}$, from (16) and (13) follows that

$$\frac{d}{dx} \widetilde{P}_n^{(s)}(x) = \frac{n(n + 2s)}{1 + 2s} \widetilde{P}_{n-1}^{(s+1)}(x) \tag{17}$$

$$\frac{d^2}{d^2x} \widetilde{P}_n^{(s)}(x) = \frac{n(n - 1)(n + 2s)(n + 2s + 1)}{(1 + 2s)(3 + 2s)} \widetilde{P}_{n-2}^{(s+2)}(x). \tag{18}$$

$P_n^{(s)}(x)$ satisfies the following differential equation:

$$(1-x^2)y'' - (2s+1)xy' + n(n+2s)y = 0. \tag{19}$$

The zeros of ultraspherical polynomials have important and well studied properties. The literature on the subject is extensive and we will not attempt to survey it. We refer to [E] and the references cited there.

The following properties are well known, and are shared also by other systems of orthogonal polynomials.

All zeros of $P_n^{(s)}(x)$ are real and simple and lie in $[-1, 1]$. Since $\frac{d}{dx}P_n^{(s)}(x) = 2sP_{n-1}^{(s+1)}(x)$, Rolle's theorem implies that between any two zeros of $P_n^{(s)}(x)$ there is a zero of $P_{n-1}^{(s+1)}(x)$.

We will denote by $z_{n,k}(s)$, $k = 1, \ldots, n$, the zeros of $P_n^{(s)}(x)$ enumerated in increasing order. That is, $-1 < z_{n,1}(s) < \ldots < z_{n,n}(s) < 1$. When there is no risk of confusion, we will just let $z_{n,j}(s) = z_j$.

To the best of our knowledge, the best available upper bound for $z_n$ is in [ADGR].

$$z_n < \sqrt{\frac{(n-1)(n+2s-2)}{(n+s-2)(n+s-1)}}\cos\left(\frac{\pi}{n+1}\right), \quad n \geq 1. \tag{20}$$

The inequality (20) improves the following inequality due to Elbert, (see [E]).

$$z_n < \frac{\sqrt{(n-1)(n+2s+1)}}{n+s} = \sqrt{1-\left(\frac{s+1}{n+s}\right)^2}. \tag{21}$$

### 2.1. *Four useful Lemmas*

**Lemma 2.1.** *Let $z_1 \ldots z_n$ be the zeros of $P_n^{(s)}(x)$ arranged in increasing order. Then*

$$P_n^{(s)}(x) = \prod_{k=\frac{[n+1]}{2}}^{n} \frac{x^2 - z_k^2}{1 - z_k^2}, \tag{22}$$

*and*

$$\prod_{k=\frac{[n+1]}{2}}^{n} (1 - z_k^2) = \prod_{k=1}^{n}(1 - z_k) = \frac{\Gamma(s)\,\Gamma(n+2\,s)}{2^n\,\Gamma(2\,s)\,\Gamma(n+s)}. \tag{23}$$

*Furthermore*

$$P_n^{(s)}(x) \leq x^n \textit{for} \;\; x \geq z_n \tag{24}$$

*Proof.* We have already observed that the zeros of $P_n^{(s)}(x)$ are symmetric with respect to $x = 0$. When $n$ is odd, $P_n^{(s)}(x)$ vanishes also at $x = 0$. Therefore, if we let $M(n,s) = \dfrac{2^n\,\Gamma(n+s)}{\Gamma(1+n)\,\Gamma(s)}$ be the coefficient of $x^n$ in the explicit expression (10) we can factorize $P_n^{(s)}(x)$ as follows:

$$P_n^{(s)}(x) = M(n,s)\prod_{k=1}^{n}(x - z_k) = M(n,s)\begin{cases}\displaystyle\prod_{k=\frac{n}{2}}^{n}(x^2 - z_k^2) & \text{if } n \text{ is even,}\\ \displaystyle x\prod_{k=\frac{n-1}{2}}^{n}(x^2 - z_k^2) & \text{if } n \text{ is odd.}\end{cases} \tag{25}$$

Thus,

$$\widetilde{P}_n^{(s)}(x) = \frac{P_n^{(s)}(x)}{P_n^{(s)}(1)} = \begin{cases}\displaystyle\prod_{k=\frac{n}{2}}^{n}\frac{x^2 - z_k^2}{1 - z_k^2} & \text{if } n \text{ is even,}\\ \displaystyle x\prod_{k=\frac{n-1}{2}}^{n}\frac{x^2 - z_k^2}{1 - z_k^2} & \text{if } n \text{ is odd}\end{cases}$$

which is (22).

Since $\frac{x^2 - z_k^2}{1 - z_k^2} \le x^2$, (24) follows.

Let $p(n,s) = \displaystyle\prod_{k=1}^{n}(1 - z_k)$. Note that $p(n,s) = \dfrac{P_n^{(s)}(1)}{M(n,s)}$, and since $P_n^{(s)}(1)$ is as in (13),

$$p(n,s) = \frac{\Gamma(s)\,\Gamma(n+2\,s)}{2^n\,\Gamma(2\,s)\,\Gamma(n+s)}$$

as required.

The inequality (24) can also be proved from the following Lemma.

**Lemma 2.2.** *for every $n > 1$ and $s > 0$, $\widetilde{P}_n^{(s)}(x) \le x\widetilde{P}_{n-1}^{(s+1)}(x)$ in $[z_n,\ 1]$.*

*Proof.* Our key tool is the differential equation (19). That is,

$$n(n+2s)y = (2s+1)xy' - (1-x^2)y'', \tag{26}$$

where $y = P_n^{(s)}(x)$.

We divide both sides of the equation (26) by $P_n^{(s)}(1)$ and recall that, by (17) and (18), $\dfrac{d}{dx}\widetilde{P}_n^{(s)}(x) = \dfrac{n(n+2s)}{1+2s}\widetilde{P}_{n-1}^{(s+1)}(x)$ and

$$\frac{d^2}{d^2x}\widetilde{P}_n^{(s)}(x) = \frac{n(n-1)(n+2s)(n+2s+1)}{(1+2s)(3+2s)}\widetilde{P}_{n-2}^{(s+2)}(x).$$

We obtain the following three term relation.

$$\frac{(n-1)(n+2s+1)}{(1+2s)(3+2s)}(1-x^2)\widetilde{P}_{n-2}^{(s+2)}(x) - x\widetilde{P}_{n-1}^{(s+1)}(x) + \widetilde{P}_n^{(s)}(x) = 0,$$

and since $\widetilde{P}_{n-2}^{(s+2)}(x)$ is positive in $[z_n,\ 1]$, we gather

$$\widetilde{P}_n^{(s)}(x) < x\widetilde{P}_{n-1}^{(s+1)}(x),$$

as required.

The following Lemma improves a Lemma in [D]

**Lemma 2.3.** *For every $0 \le |x| \le z_k$, $s > 0$ and $n \ge 2$,*

$$|\widetilde{P}_n^{(s)}(x)| \le \frac{n(n+2s)}{2s+1}\xi_{n-1}^{n-1}z_n. \tag{27}$$

*Proof.* It is well known, (see e.g. [Sz]), that the local maxima of $|P_n^{(s)}(x)|$ are increasing. The critical points of $P_n^{(s)}(x)$ are the zeros of $P_{n-1}^{(s+1)}(x)$, and hence $|P_n^{(s)}(x)|$, restricted to the interval $[0,\ z_n]$, attains its maximum at the largest zero of $P_{n-1}^{(s+1)}(x)$, which we can denote by $\xi_{n-1}$. Thus, for every $x \in [-z_n, z_n]$, $\widetilde{P}_n^{(s)}(x) \le \widetilde{P}_n^{(s)}(\xi_{n-1})$.

By the mean value theorem,

$$\widetilde{P}_n^{(s)}(z_n) - \widetilde{P}_n^{(s)}(\xi_{n-1}) = (z_n - \xi_{n-1})\frac{\partial}{\partial x}\widetilde{P}_n^{(s)}(\xi)$$

where $\xi_{n-1} < \xi < z_n$. By (17),

$$-\widetilde{P}_n^{(s)}(\xi_{n-1}) = (z_n - \xi_{n-1})\frac{n(n+2s)}{2s+1}\widetilde{P}_{n-1}^{(s+1)}(\xi)$$

and since $\widetilde{P}_{n-1}^{(s+1)}(x) \le x^{n-1}$ in $[\xi_{n-1}, 1]$ and $-\widetilde{P}_n^{(s)}(\xi_{n-1}) = |\widetilde{P}_n^{(s)}(\xi_{n-1})|$, we can infer that

$$\widetilde{P}_n^{(s)}(x) \le \xi^{n-1}(z_n - \xi_{n-1})\frac{n(n+2s)}{2s+1} < \xi_{n-1}^{n-1}z_n\frac{n(n+2s)}{2s+1},$$

as required.

The following Lemma concerns the monotonicity of ratios of Gamma functions.

**Lemma 2.4.** *a) The function*

$$x \longrightarrow \frac{\Gamma(x)\,x^y}{\Gamma(x+y)}, \quad x > 0,$$

*is decreasing when $0 < y \le 1$ and is increasing when $y > 1$. Therefore,*

$$\frac{\Gamma(x)\,x^y}{\Gamma(x+y)} \le \lim_{x\to\infty}\frac{\Gamma(x)\,x^y}{\Gamma(x+y)} = 1 \tag{28}$$

*when $y > 1$, and the inequality reverses when $y < 1$.*

*b) The function*

$$x \longrightarrow \frac{\Gamma(x)\,(x+y)^y}{\Gamma(x+y)}, \quad x > 0,$$

*is decreasing for every $y > 0$, and*

$$\frac{\Gamma(x)\, x^y}{\Gamma(x+y)} \geq \lim_{x\to\infty} \frac{\Gamma(x)\, x^y}{\Gamma(x+y)} = 1. \tag{29}$$

*Proof.* We prove only a) since the proof of b) is almost the same. Let $g_y(x) = \dfrac{\Gamma(x)\, x^y}{\Gamma(x+y)}$. When $y = 0$ and $y = 1$, then $g_y(x) \equiv 1$, so we assume either $y > 1$ or $0 < y < 1$.

To investigate the monotonicity of $g_y(x)$ we study the sign of the derivative of

$$\ln g_y(x) = y \ln x + \ln(\Gamma(x)) - \ln(\Gamma(x+y)).$$

The logarithmic derivative of $\Gamma(z)$ is

$$\frac{\Gamma'(z)}{\Gamma(z)} = \gamma - \frac{1}{z} - \sum_{m=1}^{\infty} \left( \frac{1}{z+m} - \frac{1}{m} \right)$$

where $\gamma$ is Euler's constant. Therefore,

$$\begin{aligned}(\ln g_y(x))' &= \frac{y}{x} - \sum_{m=0}^{\infty} \frac{1}{x+m} - \frac{1}{x+y+m} \\ &= y\left( \frac{1}{x} - \sum_{m=0}^{\infty} \frac{1}{(x+m)(x+m+y)} \right).\end{aligned}$$

Note that

$$\frac{1}{x} = \sum_{m=0}^{\infty} \frac{1}{x+m} - \frac{1}{x+m+1} = \sum_{m=0}^{\infty} \frac{1}{(x+m)(x+m+1)}.$$

Thus,

$$\begin{aligned}(\ln g_y(x))' &= y\left( \sum_{m=0}^{\infty} \frac{1}{(x+m)(x+m+1)} - \sum_{m=0}^{\infty} \frac{1}{(x+m)(x+m+y)} \right) \\ &= y \sum_{m=0}^{\infty} \frac{y-1}{(x+m+1)(x+m+y)}. \end{aligned} \tag{30}$$

When $y > 1$ the function in (30) is positive and when $y < 1$ it is negative. Therefore, $\ln g_y(x)$ is increasing whenever $y > 1$ and is decreasing whenever $0 \leq y < 1$, as required.

(28) follows by Stirling's formula.

## 3. Most of the Proofs

*Proof of Theorem 1.3.* We use the factorization in (22). Suppose that $n$ is even, since the proof is similar in the other case. By Hölder's inequality,

$$||\widetilde{P}_n^{(s)}||_{L^r((z_n,\ 1),\ d\mu_s)} = \left(\int_{z_n}^1 \prod_{j=\frac{n}{2}}^{n} \left(\frac{t^2-z_j^2}{1-z_j^2}\right)^r (1-t^2)^{s-\frac{1}{2}}\right)^{\frac{1}{r}}$$

$$\leq \prod_{j=\frac{n}{2}}^{n} \left(\int_{z_n}^1 \left(\frac{t^2-z_j^2}{1-z_j^2}\right)^{\frac{nr}{2}} (1-t^2)^{s-\frac{1}{2}}\right)^{\frac{2}{nr}} = \prod_{j=\frac{n}{2}}^{n} J(z_j),$$

where we have let $J(z_j) = \left(\int_{z_n}^1 \left(\frac{t^2-z_j^2}{1-z_j^2}\right)^{\frac{nr}{2}} (1-t^2)^{s-\frac{1}{2}}\,dt\right)^{\frac{2}{nr}}$.

In order to compare $J(z_j)$ with $||x^n||_{L^r(d\mu_s)}$ we let $\frac{t^2-z_j^2}{1-z_j^2} = x^2$, so that $t = \sqrt{x^2(1-z_j^2)+z_j^2}$ and $dt = \frac{x(1-z_j^2)}{\sqrt{x^2(1-z_j^2)+z_j^2}}\,dx$. Note that $x \leq \frac{x}{\sqrt{x^2(1-z_j^2)+z_j^2}} \leq 1$.

With this substitution, $(1-t^2)^{s-\frac{1}{2}} = \left((1-z_j^2)(1-x^2)\right)^{s-\frac{1}{2}}$, and

$$J(z_j)^{\frac{nr}{2}} = (1-z_j^2)^{s+\frac{1}{2}} \int_{\frac{z_n^2-z_j^2}{1-z_j^2}}^1 x^{nr}(1-x^2)^{s-\frac{1}{2}} \frac{x dx}{\sqrt{x^2(1-z_j^2)+z_j^2}}$$

$$\leq (1-z_j^2)^{s+\frac{1}{2}} \int_0^1 x^{nr}(1-x^2)^{s-\frac{1}{2}}\,dx = (1-z_j^2)^{s+\frac{1}{2}} ||x^n||^r_{L^r(d\mu_s)}$$

and

$$||\widetilde{P}_n^{(s)}||_{L^r((z_n,\ 1),\ d\mu_s)} \leq \prod_{j=1}^{\frac{n}{2}} J(z_j)$$

$$\leq ||x^n||_{L^r(d\mu_s)} \prod_{j=\frac{n}{2}}^{n} \left((1-z_j^2)^{s+\frac{1}{2}}\right)^{\frac{2}{nr}} = ||x^n||_{L^r(d\mu_s)} p(n,s)^{(s+\frac{1}{2})\frac{2}{nr}},$$

as required.

To prove the other inequality we observe that

$$\frac{x^2-z_j^2}{1-z_j^2} \geq \frac{x^2-z_n^2}{1-z_n^2}$$

whenever $j \le \frac{n}{2}$. Therefore,

$$||\widetilde{P}_n^{(s)}||_{L^r((z_n,\ 1),\ d\mu_s)} \ge \left(\int_{z_n}^1 \left(\frac{t^2 - z_n^2}{1 - z_n^2}\right)^{\frac{nr}{2}} d\mu_s(t)\right)^{\frac{1}{r}}.$$

We use again the substitution $\dfrac{t^2 - z_n^2}{1 - z_n^2} = x^2$, so that

$$||\widetilde{P}_n^{(s)}||^r_{L^r((z_n,\ 1),\ d\mu_s)} \ge (1 - z_n^2)^{s+\frac{1}{2}} \int_0^1 \frac{x^{nr+1}}{\sqrt{x^2(1 - z_n^2) + z_n^2}}(1 - x^2)^{s-\frac{1}{2}}\, dx$$

$$= (1 - z_n^2)^{s+\frac{1}{2}} \int_0^1 x^{nr}\psi(x, z_n^2)(1 - x^2)^{s-\frac{1}{2}}\, dx, \tag{31}$$

where we have let $\psi(x, t) = \dfrac{x}{\sqrt{x^2(1 - t) + t}}$.

The easy inequality $\psi(x, t) \ge x$ is not enough to prove (15). We use the elementary inequality $a^2 + b^2 - 2ab \ge 0$, with $a = \psi(x, t)^{\frac{1}{2}}$ and $b \in \mathbf{R}$, to infer that $\psi(x, t) \ge 2b(\psi(x, t))^{\frac{1}{2}} - b^2 \ge 0$ for every $b \in \mathbf{R}$. From (31) follows that

$$(1 - z_n^2)^{-(s+\frac{1}{2})}||\widetilde{P}_n^{(s)}||^r_{L^r((z_n,\ 1),\ d\mu_s)}$$

$$\ge \left(\int_0^1 2b\,(\psi(x,\ z_n^2))^{\frac{1}{2}} x^{nr}(1 - x^2)^{s-\frac{1}{2}}\, dx - b^2||x^n||_{L^r(d\mu_s)}\right). \tag{32}$$

Our next task is to choose $b$ so to maximize the function in (32).

It is easy to verify that $(\psi(x, t))^{\frac{1}{2}}$ is a convex whenever $-1 < x < 1$, and thus, by Taylor formula,

$$\psi^{\frac{1}{2}}(x, t) \ge (\psi(x, 0))^{\frac{1}{2}} + t\frac{\partial}{\partial t}(\psi(x, 0))^{\frac{1}{2}} = 1 - t\frac{1 - x^2}{2x^2},$$

and

$$||\widetilde{P}_n^{(s)}||^r_{L^r((z_n,\ 1),\ d\mu_s)}(1 - z_n^2)^{-(s+\frac{1}{2})}$$

$$\ge 2b\int_0^1 \left(1 - z_n^2\frac{1 - x^2}{2x^2}\right) x^{nr}(1 - x^2)^{s-\frac{1}{2}}\, dx - b^2||x^n||^r_{L^r(d\mu_s)}$$

$$= ||x^n||^r_{L^r(d\mu_s)}\left(2b - b^2 - b\, z_n^2\, \frac{\int_0^1 x^{nr-2}(1 - x^2)^{s-\frac{1}{2}}\, dx}{||x^n||^r_{L^r(d\mu_s)}}\right)$$

$$= b\left(2 - b - z_n^2\left(\frac{2s + 1}{2(nr - 1)}\right)\right)||x^n||^r_{L^r(d\mu_s)}.$$

The function $b \to b\left(2 - z_n^2 \left(\frac{2s+1}{2(nr-1)}\right) - b\right)$ attains its maximum when $b = \frac{1}{2}\left(2 - z_n^2 \frac{2s+1}{2(nr-1)}\right)$, and so

$$||\widetilde{P}_n^{(s)}||^r_{L^r((z_n,\,1),\,d\mu_s)} \geq (1 - z_n^2)^{s+\frac{1}{2}} \left(1 - z_n^2 \frac{2s+1}{4(nr-1)}\right)^2 ||x^n||^r_{L^r(d\mu_s)}.$$

We are left to prove that

$$c(n,s,r) = 1 - z_n^2 \frac{2s+1}{4(nr-1)}$$

is always positive. We use the upper bound for $z_n$ in (21), so to obtain

$$1 - z_n^2 \frac{2s+1}{4(nr-1)} \geq 1 - \frac{(n-1)(2s+1)(n+2s-2)\cos^2\left(\frac{\pi}{n+1}\right)}{4(nr-1)(n+s-2)(n+s-1)}$$

$$\geq 1 - \frac{(n-1)(2s+1)(n+2s-2)\cos^2\left(\frac{\pi}{n+1}\right)}{4(n-1)(n+s-2)(n+s-1)}.$$

It is easy to verify that the function above decreases with $s$ and hence,

$$1 - z_n^2 \frac{2s+1}{4(nr-1)} \geq \lim_{s\to\infty} 1 - \frac{(2s+1)(n+2s-2)\cos^2\left(\frac{\pi}{n+1}\right)}{4(n+s-2)(n+s-1)}$$

$$= 1 - \cos^2\left(\frac{\pi}{n+1}\right) = \sin^2\left(\frac{\pi}{n+1}\right)$$

and from that (15) follows.

*Proof of Proposition 1.1.* To prove that $N_2(n,s)$ decreases with $s$ we study the function $s \to \log(N_2(n,s))$; $N_2(n,s)$ is decreasing in $s$ if and only $\frac{\partial}{\partial s}\log(N_2(n,s))$ is negative.

We recall that $N_2(n,s) = \left(\frac{2^{-n}\sqrt{\pi}\Gamma(n+1)\Gamma\left(s+\frac{1}{2}\right)\Gamma(n+s)}{\Gamma\left(n+\frac{1}{2}\right)\Gamma\left(\frac{n}{2}+s\right)\Gamma\left(\frac{n}{2}+s+\frac{1}{2}\right)}\right)^{\frac{1}{2}}$.

The partial derivative of $\log(N_2(n,s))$ with respect to $s$ is

$$\frac{\partial}{\partial s}\log(N_2(n,s)$$

$$= \frac{1}{2}\sum_{m=0}^{\infty}\left(\frac{1}{\frac{n}{2}+s+m} + \frac{1}{\frac{n}{2}+s+\frac{1}{2}+m} - \frac{1}{\frac{1}{2}+s+m} - \frac{1}{n+s+m}\right)$$

$$= -\frac{n(n+1)}{2}\sum_{m=0}^{\infty}\frac{4m+2n+4s+1}{(m+n+s)(2m+2s+1)(2m+n+2s)(2m+n+2s+1)}$$

which is negative, as required.

The inequality (12) follows by Stirling formula.

*Proof of Lemma 1.2.* We use Riesz interpolation theorem. When $r \geq 2$,

$$||P_n^{(s)}||_{L^r(d\mu_s)} \leq ||P_n^{(s)}||_{L^2(d\mu_s)}^{\frac{2}{r}} ||P_n^{(s)}||_{L^\infty(d\mu_s)}^{1-\frac{2}{r}},$$

or equivalently

$$||\widetilde{P}_n^{(s)}||_{L^r(d\mu_s)} \leq ||\widetilde{P}_n^{(s)}||_{L^2(d\mu_s)}^{\frac{2}{r}}$$

since $||\widetilde{P}_n^{(s)}||_{L^\infty(d\mu_s)} = 1$. From the inequality above follows that

$$\frac{||\widetilde{P}_n^{(s)}||_{L^r(d\mu_s)}}{||x^n||_{L^r(d\mu_s)}} \leq \left(\frac{||\widetilde{P}_n^{(s)}||_{L^2(d\mu_s)}}{||x^n||_{L^2(d\mu_s)}}\right)^{\frac{2}{r}} \frac{||x^n||_{L^2(d\mu_s)}^{\frac{2}{r}}}{||x^n||_{L^r(d\mu_s)}}$$

$$= N_2(n,s)^{\frac{2}{r}} \left(\frac{\Gamma\left(n+\frac{1}{2}\right)\Gamma\left(\frac{nr}{2}+s+1\right)}{\Gamma\left(\frac{1}{2}(nr+1)\right)\Gamma(n+s+1)}\right)^{\frac{1}{r}}. \tag{33}$$

We can argue as in Lemma 1.1 to show that the function

$$n \to \frac{\Gamma\left(n+\frac{1}{2}\right)\Gamma\left(\frac{nr}{2}+s+1\right)}{\Gamma\left(\frac{1}{2}(nr+1)\right)\Gamma(n+s+1)}$$

is increasing, and is then bounded above by its limit at $n \to \infty$, which is $\left(\frac{r}{2}\right)^{s+\frac{1}{2}}$.
(28) follows from Lemma 1.1.
We are left to prove that the upper bound in (29) is actually sharp when $s = 0$. Recalling that $\lim_{s\to 0} s^{-1}P_n^{(s)}(x) = \frac{2}{n}\cos(nx)$, (see Section 6), we can see that

$$\lim_{s\to 0} ||\widetilde{P}_n^{(s)}||_{L^r(d\mu_s)} = \lim_{s\to 0} \int_{-1}^{1} |\widetilde{P}_n^{(s)}(x)|^r (1-x^2)^{s-\frac{1}{2}} dx \tag{34}$$

$$= \int_0^\pi |\cos(nt)|^r dt = \frac{\sqrt{\pi}\,\Gamma\left(\frac{r+1}{2}\right)}{\Gamma\left(1+\frac{r}{2}\right)}.$$

We have used the change of variable $x = \cos t$ in the integral in (34). Therefore,

$$N_r(n,0) = \lim_{s\to 0} \frac{||\widetilde{P}_n^{(s)}||_{L^r(d\mu_s)}}{||x^n||_{L^r(d\mu_s)}} = \left(\frac{\Gamma\left(\frac{r+1}{2}\right)\Gamma\left(\frac{nr}{2}+1\right)}{\Gamma\left(\frac{r}{2}+1\right)\Gamma\left(\frac{1}{2}(nr+1)\right)}\right)^{\frac{1}{r}}.$$

By Lemma 2.4, the function $n^{-\frac{1}{2r}}N_r(n,0)$ is increasing, and its limit is $\left(\frac{\left(\frac{r}{2}\right)^{\frac{1}{2}}\Gamma\left(\frac{r+1}{2}\right)}{\Gamma\left(\frac{r}{2}+1\right)}\right)^{\frac{1}{r}}$.

*Proof of Theorem 1.4.* We use Lemma 2.3 and interpolation. When $r \geq 2$,

$$||P_n^{(s)}||_{L^r((-z_n,\ z_n)\ d\mu_s)} \leq ||P_n^{(s)}||_{L^2(d\mu_s)}^{\frac{2}{r}} ||P_n^{(s)}||_{L^\infty(-z_n,\ z_n)}^{1-\frac{2}{r}},$$

or equivalently

$$||\widetilde{P}_n^{(s)}||_{L^r((-z_n,\ z_n),\ d\mu_s)} \leq \left(\frac{n(n+2s)}{2s+1}\xi_{n-1}^{n-1} z_n)\right)^{1-\frac{2}{r}} ||\widetilde{P}_n^{(s)}||_{L^2(d\mu_s)}^{\frac{2}{r}}.$$

From the inequality above follows that

$$\frac{||\widetilde{P}_n^{(s)}||_{L^r((-z_n,\ z_n),\ d\mu_s)}}{||x^n||_{L^r(d\mu_s)}}$$

$$\leq \left(\frac{n(n+2s)}{2s+1}\xi_{n-1}^{n-1} z_n\right)^{1-\frac{2}{r}} \left(\frac{||\widetilde{P}_n^{(s)}||_{L^2(d\mu_s)}}{||x^n||_{L^2(d\mu_s)}}\right)^{\frac{2}{r}} \frac{||x^n||_{L^2(d\mu_s)}^{\frac{2}{p}}}{||x^n||_{L^r(d\mu_s)}} \tag{35}$$

$$\leq \left(\frac{n(n+2s)}{2s+1} z_n^n\right)^{1-\frac{2}{r}} N_2(n,s)^{\frac{2}{r}} \left(\frac{\Gamma\left(n+\frac{1}{2}\right)\Gamma\left(\frac{nr}{2}+s+1\right)}{\Gamma\left(\frac{1}{2}(nr+1)\right)\Gamma(n+s+1)}\right)^{\frac{1}{r}}.$$

We use Lemma 2.4 to estimate the ratio of the Gamma functions in the inequality above.

First we apply the Lemma to the ratio

$$\frac{\Gamma\left(n+\frac{1}{2}\right)}{\Gamma\left(\frac{1}{2}(nr+1)\right)} = (n-\frac{1}{2})^{-\frac{nr}{2}+n-1}\frac{\Gamma\left(n-\frac{1}{2}\right)\left(n-\frac{1}{2}\right)^{\frac{nr}{2}-n+1}}{\Gamma\left(\frac{1}{2}(nr+1)\right)} = (n-\frac{1}{2})^{-\frac{nr}{2}+n}\frac{\Gamma(x)x^y}{\Gamma(x+y)}$$

with $x = n - \frac{1}{2}$ and $y = \frac{nr}{2} - n + 1$. Since $y > 1$, $\dfrac{\Gamma(x)x^y}{\Gamma(x+y)} < 1$.

Then we apply the Lemma to the ratio

$$\frac{\Gamma\left(\frac{nr}{2}+s+1\right)}{\Gamma(n+s+1)} = \left(\frac{nr}{2}+s+1\right)^{\frac{nr}{2}-n} \frac{\Gamma(x+y)}{(x+y)^y\,\Gamma(x)}$$

$x = n+s+1$ and $y = \frac{nr}{2} - n$. The ratio $\dfrac{\Gamma(x+y)}{(x+y)^y\,\Gamma(x)}$ is always increasing, and so it is $< 1$. Therefore,

$$\left(\frac{\Gamma\left(n+\frac{1}{2}\right)\Gamma\left(\frac{nr}{2}+s+1\right)}{\Gamma\left(\frac{1}{2}(nr+1)\right)\Gamma(n+s+1)}\right)^{\frac{1}{r}} \leq \left(\frac{\frac{nr}{2}+s+1}{n-\frac{1}{2}}\right)^{n(\frac{1}{2}-\frac{1}{r})},$$

and

$$\frac{||\widetilde{P}_n^{(s)}||_{L^r((-z_n,\ z_n),\ d\mu_s)}}{||x^n||_{L^r(d\mu_s)}} \leq \left(\frac{n(n+2s)}{2s+1}\right)^{1-\frac{2}{r}} \left(z_n^2\ \frac{\frac{nr}{2}+s+1}{n-\frac{1}{2}}\right)^{n(\frac{1}{2}-\frac{1}{r})}$$

as required.

By (20), $z_n^2 < \dfrac{(n-1)(n+2s-2)\cos^2\left(\frac{\pi}{n+1}\right)}{(n+s-2)(n+s-1)}$, and so

$$\frac{||\widetilde{P}_n^{(s)}||_{L^r((-z_n,\ z_n),\ d\mu_s)}}{||x^n||_{L^r(d\mu_s)}}$$

$$\leq \left(\frac{n(n+2s)}{2s+1}\right)^{1-\frac{2}{r}} \left(\frac{(n-1)(n+2s-2)\cos^2\left(\frac{\pi}{n+1}\right)}{(n+s-2)(n+s-1)} \times \frac{\frac{nr}{2}+s+1}{n-\frac{1}{2}}\right)^{n(\frac{1}{2}-\frac{1}{r})}.$$

When $s \to \infty$ the right hand side tends to $n^{1-\frac{2}{r}} \left(\dfrac{4(n-1)\cos^2\left(\frac{\pi}{n+1}\right)}{2n-1}\right)^{n(\frac{1}{2}-\frac{1}{r})}$. It is easy to prove that the function in parenthesis is an increasing function of $n$, and its limit is 2. This concludes the proof of the Theorem.

## References

[AAR] Andrews, G. E., Askey, R., Roy, R., *Special functions.* Encyclopedia of Mathematics and its Applications, Vol.71. Cambridge University Press, Cambridge, 1999.

[ADGR] Area, I.; Dimitrov, D. K.; Godoy, E.; Ronveaux, A., *Zeros of Gegenbauer and Hermite polynomials and connection coefficients.* Math. Comp. 73, (2004), no. 248, 1937–1951.

[D] De Carli, L., *Uniform estimates of ultraspherical polynomials of large order*, Canadian Math. Bullettin. 48 (2005), no 3, 382393.

[E] Elbert, A., *Some recent results on the zeros of Bessel functions and orthogonal polynomials*, J. Comp. Appl. Math. **133**, (2001), no. 1-2, 65–83.

[EMO] Erdelyi, A., Magnus, W., Oberhettinger,F., Tricomi, F. G., *Tables of integral transforms*, Vol. 1 – 2, McGraw-Hill book company, inc. (1954).

[St] Stein, E., *Singular integrals and differentiability properties of functions*, Princeton University Press (1970).

[Sz] Szegö, G., *Orthogonal polinomials*, AMS Colloquium publications, vol. 32, (1939).

# GENERAL MONOTONE SEQUENCES AND CONVERGENCE OF TRIGONOMETRIC SERIES

M. DYACHENKO

*Moscow State University*
*Vorobe'vy Gory, 117234, Russia*
*E-mail: dyach@mail.ru*

S. TIKHONOV

*Scuola Normale Superiore*
*Pisa Piazza dei Cavalieri 7, 56126 Italy*
*E-mail: s.tikhonov@sns.it*

In this paper we study convergence results of different types (uniform, $L_p$, almost everywhere, etc.) for one- and multidimensional trigonometric series. The sufficient conditions for these results to hold are written for the series with general monotone coefficients. The sharpness of these results is examined.

*Keywords*: Trigonometric series, Convergence, General monotone coefficients.

## 1. Introduction

In this paper we consider the series

$$\sum_{k=1}^{\infty} a_k \cos kx \tag{1}$$

and

$$\sum_{k=1}^{\infty} a_k \sin kx, \tag{2}$$

where $\{a_k\}_{k=1}^{\infty}$ is a given null sequence of complex numbers, i.e., $a_k \to 0$ as $k \to \infty$. We define by $f(x)$ and $g(x)$ the sums of the series (1) and (2) respectively at the points where the series converge.

First, it is clear [19] that if the sequence $\{a_k\}$ is of bounded variation, i.e.,

$$\sum_k |\Delta a_k| \equiv \sum_k |a_k - a_{k+1}| < \infty, \tag{3}$$

then series (1) and (2) converge for all $x$ except possibly $x = 0$, in the case of (1), and converge uniformly on any interval $[\varepsilon, 2\pi - \varepsilon]$, where $0 < \varepsilon < \pi$.

If we study the quantitative characteristic of condition (3), i.e., the speed of convergence to zero of the sums $\sum_{k=n}^{\infty} |\Delta a_k|$ or $\sum_{k=n}^{2n-1} |\Delta a_k|$, it makes sense to introduce the following classes:

$$\sum_{k=n}^{\infty} |\Delta a_k| < C\beta_n \tag{4}$$

and

$$\sum_{k=n}^{2n-1} |\Delta a_k| < C\beta_n. \tag{5}$$

Here the positive sequence $\beta = \{\beta_n\}$ is a majorant of the partial sums in (4) and (5), and $C$ is a positive number independent of $n$.

We denote[a] by $\overline{GM}^1(\beta)$ and $GM^1(\beta)$ the collections of all complex null-sequences $\{a_n\}$ satisfying (4) and (5), respectively. Such sequences are said to be called *general monotone sequences with majorant* $\beta$.

The goal of this paper is to study the almost everywhere convergence of one-dimensional series (1)-(2) and their multiple analogues with general monotone coefficients.

The paper is organized as follows. In section 2 we study $L_p$-convergence conditions for series (1)-(2). These results, in particular, imply the following: if $a \in GM^1(\beta)$, where $\beta_n = n^{-\gamma}$, then the condition $\gamma > 1 - 1/p$ is sufficient for convergence in $L_p(0, 2\pi)$ of series (1) and (2) for $1 \le p \le \infty$. However, for $a \in GM^1(\beta)$, where $\beta_n = n^{\gamma}, \gamma \ge 0$, and with some conditions of the decrease order of $a$, one can prove a.e. convergence of series (1)-(2). Sharp results of such a type are presented in section 3. In section 4, we investigate similar problem for multiple series. We conclude with Section 5, where we provide a few remarks.

## 2. Uniform and $L_p$-Convergence

It was shown in Ref. 15 that for

$$_*\beta_n = \sum_{\nu=[n/c]}^{[cn]} \frac{|a_\nu|}{\nu} \qquad \text{for some} \quad c > 1,$$

the class $GM^1(_*\beta)$ includes such sequences as quasi-monotone, $O$-regularly varying quasi-monotone[b], etc. The history of this topic can be found in e.g. Refs. 15–17.

An analogue of $GM^1(_*\beta)$ is $\overline{GM}^1(^*\beta)$, that is

$$\sum_{k=n}^{\infty} |\Delta a_k| < C(^*\beta_n) \equiv C \sum_{k=[n/c]}^{\infty} \frac{|a_k|}{k} \qquad \text{for some} \quad c > 1, \tag{6}$$

(see Ref. 7).

[a]Here 1 indicates the dimension.
[b]A sequence $\{a_n\}$ is $O$-regularly varying quasi-monotone sequence if there exists $\{\lambda_n\} \uparrow$ such that $\lambda_{2n} \le C\lambda_n$ and $\{\frac{a_n}{\lambda_n}\} \downarrow$.

First, we study the uniform convergence of the cosine and sine series.

**Theorem 2.1.** *Let $\beta = \{\beta_n\}$ be a majorant.*
**(A)** *Let $a \in \overline{GM}^1(\beta)$. Then series (1) and (2) converge for all $x$ except possibly $x = 2\pi k$, $k \in \mathbf{Z}$, in the case of (1), and converge uniformly on any interval $[\varepsilon, 2\pi-\varepsilon]$, where $0 < \varepsilon < \pi$.*
**(B)** *Let $a \in GM^1(\beta)$. If $n\beta_n = o(1)$ as $n \to \infty$, then series (1) converges uniformly on $[0, 2\pi]$ if and only if $\sum\limits_n a_n$ converges.*
**(C)** *Let $a \in GM^1(\beta)$. If $n\beta_n = o(1)$ as $n \to \infty$, then series (2) converges uniformly on $[0, 2\pi]$.*
**(D)** *If a positive sequence $a \in GM^1(_*\beta)$, then series (2) converges uniformly on $[0, 2\pi]$ if and only if $na_n = o(1)$.*

**Proof.** Item **(A)** is clear since $a$ is of bounded variation.
Let us prove item **(B)**. The part "only if" is obvious. We will show the part "if". First, it is clear that there exists $\varepsilon_n := \max\limits_{\nu \geq n}(\nu\beta_\nu) \to 0$ as $n \to \infty$.

Let $\nu_n$ be a non-increasing null-sequence such that $\left|\sum\limits_{j=n}^{\infty} a_j\right| \leq \nu_n$. Then we write

$$n\sum_{\nu=n}^{\infty} |\Delta a_\nu| = n\sum_{s=0}^{\infty}\sum_{\nu=2^s n}^{2^{s+1}n-1} |\Delta a_\nu| \leq Cn\sum_{s=0}^{\infty} \frac{\beta_{2^s n}2^s n}{2^s n} \leq C\varepsilon_n. \tag{7}$$

Therefore $\{a_n\}$ is of bounded variation and (1) converges on $(0, \pi]$. We denote its sum by $f(x)$. Let us now estimate $f(x) - S_n(f, x)$, where $S_n(f, x) = \sum\limits_{j=1}^{n} a_j \cos jx$. By the Abel transform, we get

$$f(x) - S_{n-1}(f, x) = \sum_{k=n}^{\infty} \Delta a_k D_k^*(x) - a_n D_{n-1}^*(x) =: I_1 + I_2, \qquad D_n^*(x) = \sum_{j=1}^{n} \cos jx.$$

Then, by (7),

$$|I_2| = |a_n D_{n-1}^*(x)| \leq n|a_n| \leq n\sum_{\nu=n}^{\infty} |\Delta a_\nu| \leq C\varepsilon_n.$$

To estimate $I_1$, for fixed $x \in (0, \pi]$ we define $l \in \mathbf{N}$ such that $x \in (\frac{\pi}{l+1}, \frac{\pi}{l}]$.

If $l \leq n$, then the known estimate $\left|1/2 + \sum\limits_{\nu=1}^{n} \cos \nu x\right| \leq \frac{1}{2|\sin x/2|}$ and (7) imply

$$|I_1| \leq \frac{C}{x}\sum_{k=n}^{\infty} |\Delta a_k| \leq Cl\frac{\varepsilon_n}{n} \leq C\varepsilon_n. \tag{8}$$

If $l > n$, then we decompose $I_1$ as $I_1 = \sum_{k=n}^{l-1} + \sum_{k=l}^{\infty}$. Similar to (8), we get

$$\left| \sum_{k=l}^{\infty} \triangle a_k D_k^*(x) \right| \le Cl \sum_{k=l}^{\infty} |\Delta a_k| \le C\varepsilon_l \le C\varepsilon_n.$$

Further we write

$$\left| \sum_{k=n}^{l-1} \triangle a_k D_k^*(x) \right| \le \left| \sum_{k=n}^{l-1} k \triangle a_k \right| + \sum_{k=n}^{l-1} |\Delta a_k| \, |D_k^*(x) - k| =: K + L.$$

Since

$$\sum_{k=n}^{l-1} k \triangle a_k = \sum_{k=n}^{l-1} a_k + (n-1)a_n - (l-1)a_l, \tag{9}$$

then $K \le 2\nu_n + C\varepsilon_n$.

From $|D_k^*(x) - k| \le k^2 x$, we get

$$\begin{aligned} L \le x \sum_{k=n}^{l-1} k^2 |\Delta a_k| &\le \frac{C}{l} \left( \sum_{m=n}^{l} m \sum_{j=m}^{l} |\Delta a_j| + n^2 \sum_{j=n}^{l} |\Delta a_j| \right) \\ &\le \frac{C}{l} \left( \sum_{m=n}^{l} \varepsilon_m + n\varepsilon_n \right) \le C\varepsilon_n. \end{aligned} \tag{10}$$

Collecting all estimates, we obtain $|f(x) - S_n(f,x)| \le C\Big(\varepsilon_n + \nu_n\Big)$. This gives the uniform convergence of (1).

Items **(C)** and **(D)** are known [13, 15, 18]. The proof is now complete.

**Remark 2.1.** If $n\beta_n \ne o(1)$ and $\beta_n = o(1)$ as $n \to \infty$, then there exists $a \in GM^1(\beta)$ such that series (2) converges uniformly, although $na_n \ne o(1)$ as $n \to \infty$.

Indeed, there exists a sequence $\{n_k\}_{k \in \mathbf{N}}$ such that $n_k \beta_{n_k} \ge C > 0$ for any $k$ and $\beta_{n_{k+1}} < \beta_{n_k}/2$. Hence $\sum_{k\in\mathbf{N}} \beta_{n_k} < \infty$. We define $a_m := \beta_{n_k}$ if $m = n_k$ and $a_m := 0$ otherwise. Then $a \in GM^1(\beta)$ and series (2) is uniformly convergent.

Let us now write [8] the Hardy-Littlewood-type theorem on $L_p$-convergence of series $f(x) = \sum_{k=1}^{\infty} a_k \cos kx$ and $g(x) = \sum_{k=1}^{\infty} a_k \sin kx$.

**Theorem 2.2 (Dyachenko-Tikhonov, 2008).** *Let* $1 < p < \infty$.

***(A)*** *Suppose* $a \in \overline{GM}^1(\beta)$; *then* $\sum_{n=1}^{\infty} \beta_n^p n^{p-2} < \infty \Longrightarrow f(\text{or } g) \in L_p$.

***(B)*** *If a positive sequence* $a = \{a_n\} \in \overline{GM}^1({}_*\beta)$; *then* $f(\text{or } g) \in L_p \iff \sum_{n=1}^{\infty} a_n^p n^{p-2} < \infty$.

Finally, we remark that $\sum_n |\Delta a_n| \ln n < \infty$ implies [19] $f(\text{or } g) \in L_1$. Therefore, if $a \in \overline{GM}^1(\beta)$; then $\sum_{n=1}^{\infty} \beta_n/n < \infty \Longrightarrow f(\text{or } g) \in L_1$.

### 3. Convergence Almost Everywhere: One-Dimensional Series

First, we remark that $GM^1(\beta) \equiv \overline{GM}^1(\beta)$ for $\beta_n \le n^{-\gamma}$, $\gamma > 0$. Then from the results of the previous section we have the following corollary.

**Corollary 3.1.** *If $a \in GM^1(\beta)$, where $\beta_n \le n^{-\gamma}$, then the condition $\gamma > 1 - 1/p$ is sufficient for convergence in $L_p(0, 2\pi)$ of series (1) and (2) for $1 \le p \le \infty$.*

The next theorem shows that if we assume $a \in GM^1(\beta)$ with $\beta_n \le n^{\gamma}$, $\gamma > 0$, one can expect convergence almost everywhere.

For simplicity we assume that majorants $\beta = \{\beta_n\}$ and $\gamma = \{\gamma_n\}$ satisfy the following conditions:

$$\gamma_n \asymp \gamma_m \quad \text{and} \quad \beta_n \asymp \beta_m \quad \text{for} \quad n \le m \le 2n. \tag{11}$$

Here, $\gamma_n \asymp \gamma_m$ means that $C_1\gamma_n \le \gamma_m \le C_2\gamma_n$.

Since also we use further the sequence $\gamma$ as a majorant for the null-sequence $a = \{a_n\}$, i.e., $a_n = O(\gamma_n)$, it is natural to assume that $\gamma$ is monotone decreasing.

**Theorem 3.1.** **(A)** *Let $a \in GM^1(\beta)$ and let $a_n = O(\gamma_n)$. If*

$$\sum_{n=1}^{\infty} \frac{\beta_n \gamma_n}{n} < \infty, \tag{12}$$

*then series (1) and (2) converge almost everywhere.*
**(B)** *For any majorants $\beta = \{\beta_n\}$ and $\gamma = \{\gamma_n\}$ satisfying*

$$\sum_{n=1}^{\infty} \frac{\beta_n \gamma_n}{n} = \infty \tag{13}$$

*and*

$$\beta_n \le C n \gamma_n, \tag{14}$$

*there exist almost everywhere divergent series (1) and (2) with coefficients $a \in GM^1(\beta)$ and $a_n = O(\gamma_n)$.*

**Proof. (A).** Using the Abel transform, we have for the cosine series ($a_0 = 0$)[c]

$$S_N(f, x) = \sum_{n=0}^{N} a_n \cos nx = \frac{1}{2\sin x/2}\left( \sin x/2 \sum_{n=0}^{N-1} \Delta a_n \cos nx + \right. \tag{15}$$

$$\left. + \cos x/2 \sum_{n=0}^{N-1} \Delta a_n \sin nx + a_N \sin\big((N + 1/2)x\big) \right).$$

Now we note that for $k = 1, 2, 3, \ldots$

$$\sum_{n=2^{k-1}}^{2^k-1} |\Delta a_n|^2 \le \max_{2^{k-1} \le n \le 2^k} 2|a_n| \sum_{n=2^{k-1}}^{2^k-1} |\Delta a_n| \le C\beta_{2^k}\gamma_{2^k} \le C \sum_{n=2^{k-1}}^{2^k-1} \frac{\beta_n \gamma_n}{n},$$

---

[c] In the case of $a_0/2 + \sum_{n=1}^{\infty} a_n \cos nx$, $a_0 \ne 0$ the proof is similar.

and therefore, (12) implies

$$\sum_{n=0}^{\infty} |\Delta a_n|^2 < \infty. \tag{16}$$

Therefore, using representation (15), the Carleson theorem and condition (16) imply convergence almost everywhere of series $\sum\limits_{n=0}^{\infty} a_n \cos nx$.

Let us prove **(B)**. We consider

$$b_n := \gamma_n r_n \quad \text{for} \quad n = 0, 1, 2, ...,$$

where $r_n \in \{-1; 0; 1\}$.

Since by (14) one has $\beta_n/(C\gamma_n) \leq n$, we can assume that

$$\sharp\Big\{r_n \neq 0 : n \in [2^{k-1}, 2^k - 1]\Big\} = \left[\frac{\beta_{2^{k-1}}}{C\gamma_{2^{k-1}}}\right],$$

where $\sharp M$ is the number of elements in $M$ and $[d]$ is the integer part of $d$.

Hence we have

$$\sum_{k=1}^{\infty} b_n^2 = \sum_{k=1}^{\infty} \sum_{n=2^{k-1}}^{2^k-1} b_n^2 \geq C \sum_{k=1}^{\infty} \gamma_{2^k}^2 \frac{\beta_{2^k}}{\gamma_{2^k}} \geq C \sum_{k=1}^{\infty} \frac{\gamma_k \beta_k}{k},$$

and therefore, by (13),

$$\sum_{n=0}^{\infty} b_n^2 = \infty.$$

By Zygmund's theorem (see for example, Ref. 1), one can choose numbers $r_n = \pm 1$ such that the series

$$\sum_{n=0}^{\infty} b_n \cos nx$$

will be divergent almost everywhere.

Now we check that the sequence $\{b_n\}_{n=0,1,2,...} \in GM^1(\beta)$. Indeed, for any $k \geq 1$ we write

$$\sum_{n=2^{k-1}}^{2^k-1} |\Delta b_n| \leq 2 \sum_{n=2^{k-1}}^{2^k} b_n \leq C\gamma_{2^k} \left[\frac{\beta_{2^{k-1}}}{\gamma_{2^{k-1}}}\right] \leq C\beta_{2^k}.$$

This is equivalent to $\sum\limits_{n=k}^{2k-1} |\Delta b_n| \leq C\beta_k$, that is, $\{b_n\} \in GM^1(\beta)$.

Finally, we note that clearly the condition $b_n = O(\gamma_n)$ is satisfied.

The proof for series (2) is similar. The proof is now complete.

**Remark 3.1.** Item **(A)** of Theorem 3.1 is sharp in the sense that in general the series (1) and (2) do not converge everywhere.

To prove this, we construct the series with all required conditions on coefficients and which diverges at $x_0 = \frac{\pi}{2}$. We assume that

$$\gamma_n := n^{-\delta}, \qquad a_n := \gamma_n r_n, \qquad r_n \in \{0,1\}, \qquad \delta \in \left(0, \frac{1}{2}\right),$$

where

$$\sharp\left\{r_n = 1 : n \in [2^{k-1}, 2^k - 1]\right\} = [2^{\delta(k-1)}], \qquad k > 5$$

and $r_n = 1$ only for $n = 4m$, $m \in \mathbf{N}$.

Then for this sequence we have $\{a_n\} \in GM^1(\beta)$ with $\beta_n = n^\gamma$ for any $\gamma > 0$. Hence, by Theorem 3.1 **(A)**, the series $\sum\limits_{n=0}^{\infty} a_n \cos nx$ converges almost everywhere.

On the other hand, for any $k > 5$ we get

$$\sum_{n=2^{k-1}}^{2^k-1} a_n \cos n\frac{\pi}{2} = \sum_{n=2^{k-1}}^{2^k-1} a_n \geq 2^{-\delta k} 2^{\delta(k-1)-1} = 2^{-\delta-1}.$$

and thus the series $\sum\limits_{n=0}^{\infty} a_n \cos nx$ diverges at $\frac{\pi}{2}$.

**Remark 3.2.** Condition (14) in item **(B)** of Theorem 3.1 is natural. Indeed, for any $a \in GM(\beta)$, the minimal majorant in (5) is $\overline{\beta}_n = \sum\limits_{k=n}^{2n-1} |\Delta a_k|$. Then

$$\overline{\beta}_n \leq 2 \sum_{k=n}^{2n} |a_k| \leq C \sum_{k=n}^{2n} \gamma_k \leq C n \gamma_n.$$

## 4. Convergence Almost Everywhere: Multiple Series

Let us now consider double trigonometric series

$$\sum_{m=1}^{\infty} \sum_{n=1}^{\infty} a_{m,n} \cos mx \cos ny, \tag{17}$$

$$\sum_{m=1}^{\infty} \sum_{n=1}^{\infty} a_{m,n} \sin mx \sin ny, \tag{18}$$

$$\sum_{m=1}^{\infty} \sum_{n=1}^{\infty} a_{m,n} e^{imx} e^{iny}. \tag{19}$$

We will suppose that

$$a_{m,n} \longrightarrow 0 \quad \text{as} \quad m + n \to \infty. \tag{20}$$

We denote

$$\triangle^{11} a_{m,n} := \triangle^{10}(\triangle^{01} a_{m,n}) = \triangle^{01}(\triangle^{10} a_{m,n}),$$

where $\triangle^{10} a_{m,n} = a_{m,n} - a_{m+1,n}$ and $\triangle^{01} a_{m,n} = a_{m,n} - a_{m,n+1}$.

Now we introduce the general monotone sequences in two-dimensional situation [7].

Let $\beta = \{\beta_{l,d}\}_{l,d\in\mathbf{N}}$ be a non-negative sequence. We say that a sequence $a = \{a_{m,n}\}_{m,n\in\mathbf{N}}$ satisfies *the* $\overline{GM}^2(\beta)$*-condition* or *the* $GM^2(\beta)$*-condition* if condition (20) holds and

$$\sum_{m=l}^{\infty}\sum_{n=d}^{\infty}|\Delta^{11}a_{m,n}| \le C\beta_{l,d}$$

or

$$\sum_{m=l}^{2l-1}\sum_{n=d}^{2d-1}|\Delta^{11}a_{m,n}| \le C\beta_{l,d},$$

respectively.

Similar to the sequences ${}_*\beta_n$ and ${}^*\beta_n$ in section 2, we define the double sequences

$${}_*\beta_{l,d} = \sum_{m=[l/c]}^{[cl]}\sum_{n=[d/c]}^{[cd]}\frac{|a_{m,n}|}{mn} \qquad \text{for some} \quad c>1$$

and

$${}^*\beta_{l,d} = \sum_{m=[l/c]}^{\infty}\sum_{n=[d/c]}^{\infty}\frac{|a_{m,n}|}{mn} \qquad \text{for some} \quad c>1.$$

**Theorem 4.1.** *Let* $1<p<\infty$*, and let* $a=\{a_{m,n}\}_{m,n\in\mathbf{N}} \in \overline{GM}^2(\beta)$*.*
**(A)** *If* $\beta = \{\beta_{m,n}\}_{m,n\in\mathbf{N}}$ *satisfies*

$$I(\beta) := \left(\sum_{l=1}^{\infty}\sum_{d=1}^{\infty}\beta_{l,d}^p(ld)^{p-2}\right)^{\frac{1}{p}} < \infty, \tag{21}$$

*then* $\|f_i(x,y)\|_p \le C\,I(\beta)$*,* $(i=1,2,3)$*.*
**(B)** *If a positive* $a=\{a_{m,n}\}_{m,n\in\mathbf{N}} \in \overline{GM}^2({}^*\beta)$*, then* $\|f_i(x,y)\|_p \asymp I(a)$*,* $(i=1,2,3)$*.*

Part **(A)** was verified in Ref. 8. Part **(B)** was also proved in Ref. 8 for the series with positive coefficients $a=\{a_{m,n}\}_{m,n\in\mathbf{N}} \in \overline{GM}^2(\beta)$ with

$$\beta = \left\{\beta_{l,d} := |a_{l,d}| + \sum_{m=l+1}^{\infty}\frac{|a_{m,d}|}{m} + \sum_{n=d+1}^{\infty}\frac{|a_{l,n}|}{n} + \sum_{m=l+1}^{\infty}\sum_{n=d+1}^{\infty}\frac{|a_{m,n}|}{mn}\right\}.$$

Careful analysis of this proof gives similar arguments for the more general class $\overline{GM}^2({}^*\beta)$.

Therefore, like in Corollary 3.1, if $a\in\overline{GM}^2(\beta)=GM^2(\beta)$ with $\beta_{l,d}\le(ld)^{-\gamma}$, condition $\gamma>1-1/p$ implies convergence in $L_p$ of series (17)-(19) for $1\le p\le\infty$.

Let us now write an a.e. convergence result for double series. We will consider the following convergence methods:

- Square: $\sum_{m=-\infty}^{\infty}\sum_{n=-\infty}^{\infty}$ means $\lim\limits_{K\to\infty}\sum_{|m|\le K}\sum_{|n|\le K}$,
- Rectangle (unrestricted): $\sum_{m=-\infty}^{\infty}\sum_{n=-\infty}^{\infty}$ means $\lim\limits_{\min(M,N)\to\infty}\sum_{|m|\le M}$ $\sum_{|n|\le N}$.

We again assume that majorants $\{\beta_n\}$ and $\{\gamma_n\}$ satisfy (11) and $\{\gamma_n\}$ is monotone decreasing.

**Theorem 4.2.** *Let $\beta^* = \{\beta^*_{m,n} \equiv \beta_m\beta_n\}_{m,n\in\mathbf{N}}$ and $\gamma^* = \{\gamma^*_{m,n} \equiv \gamma_m\gamma_n\}_{m,n\in\mathbf{N}}$.*
**(A)** *Let $a = \{a_{m,n}\}_{m,n\in\mathbf{N}} \in GM^2(\beta^*)$ and $|a_{m,n}| \le C\gamma^*_{m,n}$. Let also for any fixed $m\in\mathbf{N}$, $\{a_{m,n}\}_{n\in\mathbf{N}} \in GM^1(\{\beta_n\gamma_m\}_{n\in\mathbf{N}})$ and for any fixed $n\in\mathbf{N}$, $\{a_{m,n}\}_{m\in\mathbf{N}} \in GM^1(\{\beta_m\gamma_n\}_{m\in\mathbf{N}})$.*
**($\mathbf{A}_1$)** *If*

$$\sum_{n=1}^{\infty}\frac{\beta_n\gamma_n}{n} < \infty, \tag{22}$$

*then series (17)-(19) square converge almost everywhere in $\mathbf{T}_2 = \{[0,2\pi]\times[0,2\pi]\}$.*
**($\mathbf{A}_2$)** *If*

$$\sum_{n=1}^{\infty}\frac{\beta_n\gamma_n}{n}\ln n < \infty, \tag{23}$$

*then series (17)-(19) rectangular converge almost everywhere in $\mathbf{T}_2$.*
**(B)** *For any majorants $\beta = \{\beta_n\}$ and $\gamma = \{\gamma_n\}$ satisfying (13) and (14), there exist almost everywhere square divergent series (17)-(19) with coefficients $a = \{a_{m,n}\}_{m,n\in\mathbf{N}} \in GM(\beta^*)$ and $|a_{m,n}| \le C\gamma^*_{m,n}$.*

**Example 4.1.** Let $\beta_n = n^{\alpha}$ and $\gamma_n = n^{-\delta}$. Then for $\alpha\in[0,\frac{1}{2})$ and $\delta\in(\alpha,\frac{1}{2}]$ all conditions (22)-(23) hold and therefore series (17)-(19) rectangular converge almost everywhere in $\mathbf{T}_2$. If $\delta > \frac{1}{2}$, then this fact is evident (since $\sum_{m,n} a^2_{m,n}\ln n\ln m < \infty$). If $\alpha = \delta$, generally the result does not hold and series (17)-(19) could be a. e. square divergent.

**Proof.** Let us show part ($\mathbf{A}_1$). First we write for $\nu,\mu > 0$

$$\sum_{k=2^{\nu-1}-1}^{2^{\nu}-2}\sum_{n=2^{\mu-1}-1}^{2^{\mu}-2}\left(\Delta^{1,1}a_{k,n}\right)^2 \le C\max_{\substack{2^{\nu-1}-1\le k\le 2^{\nu}-1\\ 2^{\mu-1}-1\le n\le 2^{\mu}-1}}|a_{k,n}|\sum_{k=2^{\nu-1}-1}^{2^{\nu}-2}\sum_{n=2^{\mu-1}-1}^{2^{\mu}-2}|\Delta^{1,1}a_{k,n}|$$
$$\le C\gamma^*_{2^{\nu},2^{\mu}}\beta^*_{2^{\nu},2^{\mu}} \le C\sum_{k=2^{\nu-1}-1}^{2^{\nu}-2}\frac{\beta_k\gamma_k}{k}\sum_{n=2^{\mu-1}-1}^{2^{\mu}-2}\frac{\beta_n\gamma_n}{n},$$

and, therefore, (22) implies

$$\sum_{k=0}^{\infty}\sum_{n=0}^{\infty}\left(\Delta a_{k,n}\right)^2 < \infty.$$

By the theorem of Fefferman [9], Sjölin [10], and Tevzadze [14], this gives almost everywhere square convergence of the series

$$\sum_{k=0}^{\infty}\sum_{n=0}^{\infty} \Delta a_{k,n} \cos kx \cos ny \tag{24}$$

or any similar cosine-sine or sine-sine series.

We now consider the square partial sums of (17). By the Abel transform, one has ($a_{0,n} = a_{k,0} = 0$)

$$\begin{aligned}
S_{N,N} &= \sum_{k=0}^{N}\sum_{n=0}^{N} a_{k,n} \cos kx \cos ky \\
&= \frac{1}{4\sin x/2 \sin y/2} \times \left( \sum_{k=0}^{N-1}\sum_{n=0}^{N-1} \Delta^{1,1} a_{k,n} \sin(k+1/2)x \sin(n+1/2)y \right. \\
&+ \left. \sin(N+1/2)x \sum_{n=0}^{N} a_{N,n} \cos ny + \sin(N+1/2)y \sum_{k=0}^{N-1} \Delta^{1,0} a_{k,N} \sin(k+1/2)x \right) \\
&=: \frac{1}{4\sin x/2 \sin y/2} \times (J_1 + \sin(N+1/2)xJ_2 + \sin(N+1/2)yJ_3).
\end{aligned}$$

We note that $J_1$ is a linear combination of cubic partial sums of series (24) with coefficients $\cos x/2 \sin y/2$ or similar ones.

Then it is sufficient to show that the remainder terms tend to zero a.e. Again, using the Abel transform, for any $y \in (0, 2\pi)$ we get

$$\begin{aligned}
|J_2| &= \frac{1}{2\sin y/2} \left| \sum_{n=0}^{N-1} \Delta^{0,1} a_{N,n} \sin(n+1/2)y + \sin(N+1/2)y a_{N,N} \right| \\
&\le C(y) \left( \sum_{n=0}^{N-1} |\Delta^{0,1} a_{N,n}| + |a_{N,N}| \right) \le C(y) \left( \sum_{n=1}^{N} \frac{\beta_n \gamma_N}{n} + o(1) \right).
\end{aligned}$$

as $N \to \infty$. Let us now prove that (22) implies

$$\sum_{n=1}^{N} \frac{\beta_n}{n} = o(1/\gamma_N) \qquad \text{as} \qquad N \to \infty. \tag{25}$$

Let $\varepsilon > 0$. Fix integer $k_0$ such that

$$\sum_{n=k_0+1}^{\infty} \frac{\beta_n \gamma_n}{n} < \frac{\varepsilon}{2}.$$

We now select $N_0$ as follows:

$$\gamma_{N_0} \sum_{n=1}^{k_0} \frac{\beta_n}{n} < \frac{\varepsilon}{2}.$$

Therefore, for any $N \geq \max(N_0, k_0 + 1)$ we get

$$\gamma_N \sum_{n=1}^{N} \frac{\beta_n}{n} \leq \gamma_{N_0} \sum_{n=1}^{k_0} \frac{\beta_n}{n} + \sum_{n=k_0+1}^{N} \frac{\beta_n \gamma_n}{n} < \varepsilon.$$

Thus (25) holds. This gives $|J_2| = o(1)$ as $N \to \infty$.

Similarly,

$$|J_3| \leq \sum_{k=0}^{N-1} |\Delta^{1,0} a_{k,N}| \leq C \sum_{n=1}^{N} \frac{\beta_n \gamma_N}{n} = o(1)$$

as $N \to \infty$.

Part $(\mathbf{A}_2)$ one can prove similarly. Indeed, (23) implies

$$\sum_{k=0}^{\infty} \sum_{n=0}^{\infty} (\Delta a_{k,n})^2 \Big( \ln \min\left(k+2, n+2\right) \Big)^2 < \infty.$$

By Theorem 7.2 of Sjölin's paper Ref. 10, this implies almost everywhere rectangular convergence of the series

$$\sum_{k=0}^{\infty} \sum_{n=0}^{\infty} \Delta a_{k,n} \cos kx \cos ny.$$

The rest of the proof is verbatim.

To prove part **(B)**, we use the example from Theorem 3.1 **(B)**. Indeed, let us write the series which was constructed in the proof of Theorem 3.1 **(B)** as

$$\sum_{n=1}^{\infty} a'_n \cos nx. \tag{26}$$

Let $S_k(x)$ be its $n$-th partial sum. Then the double series

$$\sum_{n=1}^{\infty} a'_n \cos nx \cos y \tag{27}$$

satisfies all conditions of part **(B)**.

First, we remark that all required estimates for $\{a_{n,m}\}$ and $\{\Delta a_{n,m}\}$ are fulfilled. We also note that the $(k,k)$-th (square) partial sums of (27) coincide with $S_k(x) \cos y$. This gives the a.e. square divergence of (27) in $\mathbf{T}_2$.

The proof is now complete.

## 5. Concluding Remarks

**Remark 5.1.** As C. Fefferman [11] proved, the double Fourier series of a continuous function may fail to converge in the unrestricted rectangular sense at each point of a set of positive measure. Thus it is interesting to show different types of convergence (a.e., $L_p$) for functions whose Fourier coefficients exhibit a certain amount of good behavior.

Results of Theorems 3.1 and 4.2 supplement the results on Móricz and Waterman [12] on double series with some conditions on $\{\Delta^{1,0}a_{m,n}\}$, $\{\Delta^{0,1}a_{m,n}\}$, and $\{\Delta^{1,1}a_{m,n}\}$. See also Ref. 2.

**Remark 5.2.** We now present some sufficient conditions for $u$-convergence of multiple series. First we recall the

**Definition.** Let $U \subset \mathbf{Z}^n$. Then we say that $U \in A$ if $\mathbf{k} \in U$ implies $\prod_{j=1}^{n} [-|k_j|, |k_j|] \cap \mathbf{Z}^n \subseteq U$. We also say that the numerical series $\sum_{\mathbf{m} \in \mathbf{Z}^n} c_{\mathbf{m}}$ *u-converges* to the number $\alpha$ if for any $\varepsilon > 0$ there exists a number $M$ such that for every $U \in A$ for which $\left\{\mathbf{m} \in \mathbf{Z}^n : |\mathbf{m}| \leq M\right\} \subseteq U$, we have $\left|\sum_{\mathbf{m} \in U} c_{\mathbf{m}} - \alpha\right| < \varepsilon$. Similarly, we define $u$-convergence in $L_p$.

We note that $u$-convergence implies convergence over rectangles, over spheres, over hyperbolic crosses, etc. (see for details [3–6] and references therein).

**Corollary 5.1.** **(A)** *Under the conditions of Theorem 4.1 (A), series (17)-(19) $u$-converge in $L_p$ for $4/3 < p < \infty$.*

**(B)** *Suppose $1 < p \leq 4/3$. Then there exist functions $f_1$, $f_2$ and $f_3 \in L_p$ such that*

1. *their Fourier series are of the form (17)-(19),*
2. *$\{a_{m,n}\}_{m,n \in \mathbf{N}} \in \overline{GM}^2(\beta)$, where $\beta = \{\beta_{m,n}\}_{m,n \in \mathbf{N}}$ satisfies (21),*
3. *(17)-(19) are $u$-divergent in $L_p$.*

*Proof.* Indeed, if

$$I(\beta) := \left( \sum_{l=l}^{\infty} \sum_{d=1}^{\infty} \beta_{l,d}^p (ld)^{p-2} \right)^{\frac{1}{p}} < \infty,$$

then denoting

$$a'_{m,n} := \sum_{l=m}^{\infty} \sum_{d=n}^{\infty} |\Delta^{11} a_{l,d}|$$

and

$$a''_{m,n} := a'_{m,n} - a_{m,n}$$

for any $m, n \geq 1$, we get

$$\Delta^{11} a'_{m,n} = |\Delta^{11} a_{m,n}| \geq 0 \qquad \text{and} \qquad \Delta^{11} a''_{mn} = |\Delta^{11} a_{m,n}| - \Delta^{11} a_{m,n} \geq 0.$$

Hence, both $\mathbf{a}'$ and $\mathbf{a}''$ satisfy the following condition:

$$\Delta^{11} a_{m,n} \geq 0$$

and therefore satisfy the condition

$$a_{m_1,n_1} \leq a_{m,n} \quad \text{for} \quad m_1 \geq m \quad \text{and} \quad n_1 \geq n.$$

Further, $I(\{a'\}) < \infty$ implies $I(\{a\}) < \infty$, and then we get $I(\{a''\}) < \infty$. Now we can apply Theorem 1 from Ref. 5. Thus, for $\frac{4}{3} < p < \infty$ series (17)-(19) $u$-converge in $L_p$.

The second part follows from the proof of Theorem 2 of the same paper Ref. 5. The proof is now complete.

**Remark 5.3.** For $n$-dimensional case ($n \geq 2$) the critical value for $p$ is $\frac{2n}{n+1}$.

## Acknowledgments

The present work was supported by the Russian Foundation for Fundamental Research (grant 06-01-00268) and the Leading Scientific Schools (grant NSH-2787.2008.1).

## References

[1] N. K. Bary, *A treatise on trigonometric series.* Vol. I, II. (Pergamon Press, 1964).
[2] C. P. Chen, P. H. Hsieh, Pointwise convergence of double trigonometric series. With a note by Ferenc Móricz, *J. Math. Anal. Appl.*, **172**, 582 (1993).
[3] M. I. Dyachenko, Norms of Dirichlet kernels and some other trigonometric polynomials in $L_p$-spaces, *Mat. Sb.* **184**, 3 (1993), translation in *Russ. Acad. Sci., Sb. Math.*, **78**, 267 (1994).
[4] M. I. Dyachenko, $u$-convergence of multiple Fourier series, *Izv. Ross. Akad. Nauk, Ser. Mat.* **59**, 129 (1995), translation in *Izv. Math.* **59**, 353 (1995).
[5] M. I. Dyachenko, $U$-convergence of Fourier series with monotone and with positive coefficients, *Mat. Zametki* **70**, 356 (2001); translation in *Math. Notes* **70**, 320 (2001).
[6] M. I. Dyachenko, *Convergence of multiple Fourier series: main results and unsolved problems* in *Fourier analysis and related topics. Banach Cent. Publ.* **56**, 37 (2002).
[7] M. Dyachenko, S. Tikhonov, Convergence of trigonometric series with general monotone coefficients, *C. R. Math. Acad. Sci. Paris* **345**, 3, 123 (2007).
[8] M. Dyachenko, S. Tikhonov, A Hardy-Littlewood theorem for multiple series, *J. of Math. Anal. Appl.* **339**, 503, (2008).
[9] C. Fefferman, On the convergence of multiple Fourier series, *Bull. Amer. Math. Soc.* **77**, 744 (1971).
[10] P. Sjölin, Convergence almost everywhere of certain singular integrals and multiple Fourier series, *Arkiv för Math.* **9**, 65 (1971).
[11] C. Fefferman, On the divergence of multiple Fourier series, *Bull. Amer. Math. Soc.* **77**, 191 (1971).
[12] F. Móricz, D. Waterman, Convergence of double Fourier series with coefficients of generalized bounded variation, *J. of Math. Anal. Appl.* **140**, 34, (1989).
[13] L. Leindler, A new extension of monotone sequences and its applications. *J. Inequal. Pure Appl. Math.* **7**, 39 (2006).
[14] N. R. Tevzadze, The convergence of the double Fourier series at a square summable function. (Russian) *Sakharth. SSR Mecn. Akad. Moambe*, **58**, 277 (1970).
[15] S. Tikhonov, Best approximation and moduli of smoothness: computation and equivalence theorems. To appear in *J. Approx. Theory.* Available online at `http://www.sciencedirect.com/science/journal/00219045`.
[16] S. Tikhonov, Trigonometric series with general monotone coefficients, *J. of Math. Anal. Appl.*, **326**, 721, (2007).

[17] S. Tikhonov, On uniform convergence of trigonometric series, *Mat. Zametki*, **81**, 304 (2007), translation in *Math. Notes*, **81**, 268 (2007).

[18] S. P. Zhou, P. Zhou, D. S. Yu, Ultimate generalization to monotonicity for uniform convergence of trigonometric series, *arXiv:math.CA*/0611805.

[19] A. Zygmund, *Trigonometric series.* Vol. I, II. 3rd edn. (Cambridge, 2002).

# USING INTEGRALS OF SQUARES OF CERTAIN REAL-VALUED SPECIAL FUNCTIONS TO PROVE THAT THE PÓLYA $\Xi^*(z)$ FUNCTION, THE FUNCTIONS $K_{iz}(a), a > 0$, AND SOME OTHER ENTIRE FUNCTIONS HAVE ONLY REAL ZEROS

GEORGE GASPER

*Department of Mathematics, Northwestern University,*
*2033 Sheridan Road, Evanston, IL 60208-2730, USA*
*george@math.northwestern.edu*
*www.math.northwestern.edu*

Analogous to the use of sums of squares of certain real-valued special functions to prove the reality of the zeros of the Bessel functions $J_\alpha(z)$ when $\alpha \geq -1$, confluent hypergeometric functions ${}_0F_1(c; z)$ when $c > 0$, Laguerre polynomials $L_n^\alpha(z)$ when $\alpha \geq -2$, Jacobi polynomials $P_n^{(\alpha,\beta)}(z)$ when $\alpha \geq -1$ and $\beta \geq -1$, and some other entire special functions considered in G. Gasper [Using sums of squares to prove that certain entire functions have only real zeros, in *Fourier Analysis: Analytic and Geometric Aspects*, W. O. Bray, P. S. Milojević and C. V. Stanojević, eds. (Marcel Dekker, Inc., 1994) pp. 171–186.], integrals of squares of certain real-valued special functions are used to prove the reality of the zeros of the Pólya $\Xi^*(z)$ function, the $K_{iz}(a)$ functions when $a > 0$, and some other entire functions.

*Keywords*: $K_{iz}(a)$ functions, Pólya $\Xi^*$ function, Riemann $\Xi$ function, reality of zeros of entire functions, integrals of squares, sums of squares, absolutely monotonic functions, convex functions, nonnegative functions, special functions, inequalities, Fourier and cosine transforms, Meijer $G$ functions, Mellin-Barnes integrals, modified Bessel functions of the third kind, continuous dual Hahn polynomials.

**Dedicated to Dan Waterman on the occasion of his 80th birthday**

## 1. Introduction

It is well-known [1] that the Riemann Hypothesis is equivalent to the statement that all of the zeros of the Riemann $\Xi(z)$ function are real. $\Xi(z)$ is an even entire function of $z$ with the integral representations

$$\Xi(z) = \int_{-\infty}^{\infty} \Phi(u)\, e^{izu}\, du = 2\int_{0}^{\infty} \Phi(u)\cos(zu)\, du, \tag{1}$$

where

$$\Phi(u) = \sum_{n=1}^{\infty}(4n^4\pi^2 e^{\frac{9}{2}u} - 6n^2\pi e^{\frac{5}{2}u})e^{-n^2\pi e^{2u}}. \tag{2}$$

In a 1926 paper Pólya [2] observed that

$$\Phi(u) \sim 8\pi^2 \cosh(\frac{9}{2}u)\, e^{-2\pi \cosh(2u)} \quad \text{as } u \to \pm\infty \tag{3}$$

and, in view of this asymptotic equivalence to $\Phi(u)$, considered the problem of determining whether or not the entire function

$$\Xi^*(z) = 16\pi^2 \int_0^\infty \cosh(\frac{9}{2}u)\, e^{-2\pi \cosh(2u)} \cos(zu)\, du \tag{4}$$

has only real zeros. Here, as is now customary, the capital letter $\Xi$ is used instead of the original lower case $\xi$. Pólya was able to prove that $\Xi^*(z)$ has only real zeros by using (in a different notation) a difference equation in $z$ for the modified Bessel function of the third kind [3, 4]

$$K_z(a) = \int_0^\infty e^{-a\cosh u} \cosh(zu)\, du, \qquad a > 0, \tag{5}$$

to prove for each $a > 0$ that $K_{iz}(a)$ has only real zeros, and then applying the identity

$$\Xi^*(z) = 4\pi^2 [K_{\frac{1}{2}iz - \frac{9}{4}}(2\pi) + K_{\frac{1}{2}iz + \frac{9}{4}}(2\pi)] \tag{6}$$

and the special case $G(z) = K_{iz/2}(2\pi), c = \frac{9}{4}$, of the lemma (derived via an infinite product representation for $G(z)$):

**Lemma.** *If $-\infty < c < \infty$ and $G(z)$ is an entire function of genus 0 or 1 that assumes real values for real $z$, has only real zeros and has at least one real zero, then the function*

$$G(z - ic) + G(z + ic)$$

*also has only real zeros.*

More generally, in a subsequent paper Pólya [5] pointed out that from the case $G(z) = K_{iz/2}(a)$ of this lemma it follows that each of the entire functions

$$\begin{aligned} F_{a,c}(z) &= K_{i(z-ic)}(a) + K_{i(z+ic)}(a) \\ &= 2\int_0^\infty \cosh(cu)\, e^{-a\cosh u} \cos(zu)\, du, \qquad a > 0,\ -\infty < c < \infty, \end{aligned} \tag{7}$$

has only real zeros. He also used a differential equation in the variable $a$ to give a proof of the reality of the zeros of $K_{iz}(a), a > 0$, that was simpler than his previous proof.

Our main aim in this paper is to show how integrals of squares of certain real-valued special functions can be used to give new proofs of the reality of the zeros of the above $\Xi^*(z), K_{iz}(a)$, and $F_{a,c}(z)$ functions. This paper is a sequel to the author's 1994 paper [6] in which he showed how sums of squares of certain real-valued special functions could be used to prove the reality of the zeros of the Bessel functions $J_\alpha(z)$ when $\alpha \geq -1$, confluent hypergeometric functions ${}_0F_1(c; z)$ when

$c > 0$ or $0 > c > -1$, Laguerre polynomials $L_n^\alpha(z)$ when $\alpha \geq -2$, Jacobi polynomials $P_n^{(\alpha,\beta)}(z)$ when $\alpha \geq -1$ and $\beta \geq -1$, and some other entire functions. Also see the applications of squares of real-valued special functions in Refs. 7–16.

## 2. Reality of the Zeros of the Functions $K_{iz}(a)$ When $a > 0$

Let $a > 0$ and $z = x+iy$, where $x$ and $y$ are real variables. First observe that, by the Meijer $G$-function representation for the product of two modified Bessel functions of the third kind (Eq. 5.6(66) in Ref. 3) and the definition of a Meijer $G$-function as a Mellin-Barnes integral (Eq. 5.3(1) in Ref. 3),

$$|K_{iz}(a)|^2 = K_{iz}(a)K_{i\bar{z}}(a) = \frac{\sqrt{\pi}}{2} G_{24}^{40}\left[a^2 \middle| \begin{matrix} 0, \frac{1}{2} \\ ix, -ix, y, -y \end{matrix}\right]$$

$$= \frac{\sqrt{\pi}}{4\pi i}\int_{c-i\infty}^{c+i\infty} \frac{\Gamma(ix-s)\Gamma(-ix-s)\Gamma(y-s)\Gamma(-y-s)}{\Gamma(-s)\Gamma(\frac{1}{2}-s)} a^{2s}\,ds \quad (8)$$

with $c < -|y|$, where the path of integration is along the upwardly oriented vertical line $\Re(s) = c$. Next note that, by Gauss' summation formula (Eq. (1.2.11) in Ref. 16),

$$\sum_{k=0}^{\infty} \frac{(y)_k (y)_k}{k!\,(y-s)_k} = {}_2F_1\left[\begin{matrix} y, y \\ y-s \end{matrix}; 1\right] = \frac{\Gamma(y-s)\Gamma(-y-s)}{\Gamma(-s)\Gamma(-s)}, \qquad \Re(s) < -y,$$

where $(y)_k = \Gamma(y+k)/\Gamma(y)$ and the series is absolutely convergent. Hence,

$$|K_{iz}(a)|^2 = \frac{\sqrt{\pi}}{4\pi i}\int_{c-i\infty}^{c+i\infty} \sum_{k=0}^{\infty} \frac{((y)_k)^2\Gamma(ix-s)\Gamma(-ix-s)\Gamma(y-s)\Gamma(-s)}{k!\,\Gamma(\frac{1}{2}-s)\Gamma(y+k-s)} a^{2s}\,ds$$

$$= [K_{ix}(a)]^2 + \frac{\sqrt{\pi}}{4\pi i}\sum_{k=1}^{\infty} \frac{((y)_k)^2}{k!}\int_{c-i\infty}^{c+i\infty} \frac{\Gamma(ix-s)\Gamma(-ix-s)\Gamma(y-s)\Gamma(-s)}{\Gamma(\frac{1}{2}-s)\Gamma(y+k-s)} a^{2s} ds$$

and, since $(y)_k = y(y+1)_{k-1}$ for $k \geq 0$,

$$|K_{iz}(a)|^2 = [K_{ix}(a)]^2 + y^2 L_a(x,y) \quad (9)$$

with

$$L_a(x,y) = \frac{\sqrt{\pi}}{4\pi i}\sum_{k=0}^{\infty} \frac{((y+1)_k)^2}{(k+1)!}\int_{c-i\infty}^{c+i\infty} \frac{\Gamma(ix-s)\Gamma(-ix-s)\Gamma(y-s)\Gamma(-s)}{\Gamma(\frac{1}{2}-s)\Gamma(y+k-s)} a^{2s} ds. \quad (10)$$

From (9) it follows that in order to prove that $K_{iz}(a)$ has only real zeros it suffices to prove that

$$L_a(x,y) > 0, \qquad -\infty < x, y < \infty. \quad (11)$$

To prove (11) from (10) we observe that

$$\int_0^1 t^{y-s-1}(1-t)^{k-1}\,dt = \frac{\Gamma(k)\Gamma(y-s)}{\Gamma(y+k-s)}, \qquad k > 0,\ \Re(s) < y,$$

by the beta integral (Eq. (1.11.8) in Ref. 16), and thus

$$\frac{1}{2\pi i}\int_{c-i\infty}^{c+i\infty}\frac{\Gamma(ix-s)\Gamma(-ix-s)\Gamma(y-s)\Gamma(-s)}{\Gamma(\frac{1}{2}-s)\Gamma(y+k-s)}a^{2s}\,ds$$

$$=\frac{1}{\Gamma(k)}\int_0^1 t^{y-1}(1-t)^{k-1}\left[\int_{c-i\infty}^{c+i\infty}\frac{\Gamma(ix-s)\Gamma(-ix-s)\Gamma(-s)}{2\pi i\,\Gamma(\frac{1}{2}-s)}\left(\frac{a^2}{t}\right)^s ds\right]dt$$

$$=\frac{2}{\sqrt{\pi}\Gamma(k)}\int_0^1 t^{y-1}(1-t)^{k-1}\left[K_{ix}\left(\frac{a}{\sqrt{t}}\right)\right]^2 dt$$

by Fubini's Theorem and (8), which gives

$$L_a(x,y)=\sum_{k=0}^{\infty}\frac{((y+1)_k)^2}{k!(k+1)!}\int_0^1 t^{y-1}(1-t)^{k-1}\left[K_{ix}\left(\frac{a}{\sqrt{t}}\right)\right]^2 dt>0 \tag{12}$$

for $-\infty<x,y<\infty$. Equations (9) and (12) can be combined to give the formula

$$|K_{iz}(a)|^2=[K_{ix}(a)]^2+y^2\int_0^1 t^{y-1}{}_2F_1\left[\begin{matrix}y+1,y+1\\2\end{matrix};1-t\right]\left[K_{ix}\left(\frac{a}{\sqrt{t}}\right)\right]^2 dt, \tag{13}$$

from which it follows that $K_{iz}(a)$ has only real zeros when $a>0$, since the integrand is clearly nonnegative.

The reality of the zeros of $K_{iz}(a), a>0$, can also be proved by taking the first and second partial derivatives of the formula (13) with respect to $y$ to obtain the formulas

$$y\frac{\partial}{\partial y}|K_{iz}(a)|^2=\int_0^1 y\frac{\partial}{\partial y}f_t(y)\left[K_{ix}\left(\frac{a}{\sqrt{t}}\right)\right]^2\frac{dt}{t} \tag{14}$$

and

$$\frac{\partial^2}{\partial y^2}|K_{iz}(a)|^2=\int_0^1 \frac{\partial^2}{\partial y^2}f_t(y)\left[K_{ix}\left(\frac{a}{\sqrt{t}}\right)\right]^2\frac{dt}{t} \tag{15}$$

with

$$f_t(y)=y^2t^y{}_2F_1\left[\begin{matrix}y+1,y+1\\2\end{matrix};1-t\right]. \tag{16}$$

In Ref. 20 Askey and the author utilized the reality of the zeros of the continuous dual Hahn polynomials (p. 331 in Ref. 21) to show that $f_t(y)$ (and a generalization of it) is an (even) absolutely monotonic function (one whose power series coefficients are nonnegative) of $y$ when $0<t<1$, which shows that

$$y\frac{\partial}{\partial y}f_t(y)\geq 0,\qquad -\infty<y<\infty,\ \ 0<t<1, \tag{17}$$

and

$$\frac{\partial^2}{\partial y^2}f_t(y)\geq 0,\qquad -\infty<y<\infty,\ \ 0<t<1. \tag{18}$$

and, hence, that the integrals in (14) and (15) are positive when $y \neq 0$. Thus, it follows from (14) that $|K_{iz}(a)|^2$ is an increasing (decreasing) even function of $y$ when $y > 0$ ($y < 0$), and it follows from (15) that $|K_{iz}(a)|^2$ is a convex even non-constant function of $y$, each of which implies that for any $x$ the function $|K_{iz}(a)|^2$ assumes its minimum value when $y = 0$ and proves that $K_{iz}(a)$ has only real zeros.

Corresponding to the above proofs via formulas (14) and (15), it should be noted that in 1913 Jensen [17] showed that each of the inequalities

$$y\frac{\partial}{\partial y}|F(x+iy)|^2 \geq 0, \quad -\infty < x, y < \infty, \tag{19}$$

and

$$\frac{\partial^2}{\partial y^2}|F(x+iy)|^2 \geq 0, \quad -\infty < x, y < \infty, \tag{20}$$

is necessary and sufficient for a real entire function $F(z) \not\equiv 0$ of genus 0 or 1 to have only real zeros (see Chapter 2 in Ref. 18 and the necessary and sufficient conditions in Ref. 19).

## 3. Reality of the Zeros of the Functions $\Xi^*(z)$ and $F_{a,c}(z)$

Because of $\Xi^*(z) = 4\pi^2 F_{2\pi,9/4}(z/2)$, it suffices to show how formula (13) can be used to prove that $F_{a,c}(z)$ has only real zeros when $a, c > 0$. Fix $a, c > 0$ and suppose that $z_0 = x_0 + iy_0$ is a zero of $F_{a,c}(z)$. Then $K_{i(x_0+i(y_0+c))}(a) = -K_{i(x_0+i(y_0-c))}(a)$ by (7) and hence

$$\begin{aligned} 0 &= |K_{i(x_0+i(y_0+c))}(a)|^2 - |K_{i(x_0+i(y_0-c))}(a)|^2 \\ &= \int_0^1 [f_t(y_0+c) - f_t(y_0-c)]\left[K_{ix_0}\left(\frac{a}{\sqrt{t}}\right)\right]^2 \frac{dt}{t} \end{aligned} \tag{21}$$

by (13) and (16). Since $f_t(y)$ is an even convex non-constant function of $y$ when $0 < t < 1$,

$$f_t(y_0+c) - f_t(y_0-c) \begin{cases} > 0 & \text{if } y_0 > 0, \\ = 0 & \text{if } y_0 = 0, \\ < 0 & \text{if } y_0 < 0, \end{cases}$$

and it follows from (21) that $y_0 = 0$ and, thus, the function $F_{a,c}(z)$ has only real zeros when $a, c > 0$.

One can also try to give another proof of the reality of the zeros of $F_{a,c}(z)$ via a formula for the function

$$\begin{aligned} |F_{a,c}(z)|^2 &= F_{a,c}(z)F_{a,c}(\bar{z}) \\ &= K_{ix-y-c}(a)K_{ix+y-c}(a) + K_{ix-y-c}(a)K_{ix+y+c}(a) \\ &\quad + K_{ix-y+c}(a)K_{ix+y-c}(a) + K_{ix-y+c}(a)K_{ix+y+c}(a) \end{aligned} \tag{22}$$

that contains integrals of nonnegative functions as in (13). By using the right-hand side of (22), Eqs. 5.6(66) and 5.3(1) in Ref. 3, the Gauss summation formula and the beta integral, it can be shown that

$$|F_{a,c}(z)|^2 = [F_{a,c}(x)]^2 + \int_0^1 f_t(y)[F_{a/\sqrt{t},\,c}(x)]^2 \frac{dt}{t} + \int_0^1 g_{t,c}(y)|K_{i(x+ic)}(a/\sqrt{t})|^2 \frac{dt}{t} \tag{23}$$

with

$$g_{t,c}(y) = y(y+2c)t^{y+c}{}_2F_1\left[\frac{y+1, y+2c+1}{2}; 1-t\right]$$
$$+ y(y-2c)t^{y-c}{}_2F_1\left[\frac{y+1, y-2c+1}{2}; 1-t\right] - 2y^2 t^y {}_2F_1\left[\frac{y+1, y+1}{2}; 1-t\right]. \tag{24}$$

The function $g_{t,c}(y)$ is clearly an even function of $c$ such that $g_{t,0}(y) = 0$ and $g_{t,c}(0) = 0$. It can be shown that $g_{t,c}(y)$ is also an even function of $y$ by applying the Euler transformation formula Eq. (1.4.2) in Ref. 16 to the hypergeometric functions in (24). Extensive analysis of $g_{t,c}(y)$ via Mathematica and generating functions of the form of Eq. 5 in Ref. 20 strongly suggest that $g_{t,c}(y)$ is nonnegative, convex, and an absolutely monotonic function of $y$ when $0 < t < 1$ and $-\infty < c < \infty$. If the nonnegativity, convexity, or absolute monotonicity of $g_{t,c}(y)$ could be proved, then (23) and its partial derivatives with respect to $y$ would give additional proofs of the reality of the zeros of $F_{a,c}(z)$.

Pólya [5] derived a theorem concerning the zeros of Fourier transforms and universal factors, and then used it to prove that his [2]

$$\Xi^{**}(z) = 8\pi \int_0^\infty \left[2\pi \cosh(\frac{9}{2}u) - 3\cosh(\frac{5}{2}u)\right] e^{-2\pi\cosh(2u)} \cos(zu)\, du, \tag{25}$$

function and the more general functions

$$\Xi_{A,B,a,b,c}(z) = \int_0^\infty [A\cosh(au) - B\cosh(bu)]e^{-c\cosh(u)} \cos(zu)\, du \tag{26}$$

with $A > B > 0, a > b > 0, c > 0$, have only real zeros. Analogous to the formulas in (13) and (23), it might be possible to derive formulas containing squares of real-valued functions that give new proofs of the reality of the zeros of the entire functions in (25) and (26), and even, perhaps, prove that some of the entire functions in Eq. (0.3) of Hejhal [22] have only real zeros.

## Acknowledgment

The author wishes to thank the referee for suggesting some improvements in the paper.

## References

[1] E. C. Titchmarsh, *The Theory of the Riemann Zeta-Function,* 2nd edn. revised by D. R. Heath-Brown (Oxford Univ. Press, Oxford and New York, 1986).

[2] G. Pólya, Bemerkung über die Integraldarstellung der Riemannschen $\zeta$-Funktion, *Acta Math.* **48**(1926), 305-317; reprinted in his *Collected Papers,* Vol. II, pp. 243–255.

[3] A. Erdelyi, *Higher Transcendental Functions,* Vols. I & II (McGraw Hill, New York, 1953).

[4] G. N. Watson, *Theory of Bessel Functions* (Cambridge Univ. Press, 1944).

[5] G. Pólya, Über trigonometrische Integrale mit nur reellen Nullstellen, *J. Reine Angew. Math.* **158**(1927), pp. 6-18; reprinted in his *Collected Papers,* Vol. II, pp. 265–275.

[6] G. Gasper, Using sums of squares to prove that certain entire functions have only real zeros, in *Fourier Analysis: Analytic and Geometric Aspects,* W. O. Bray, P. S. Milojević and C. V. Stanojević, eds. (Marcel Dekker, Inc., 1994) pp. 171–186. Preprints of this paper and some other papers by the author are available at http://www.math.northwestern.edu/~george/preprints/ .

[7] R. Askey and G. Gasper, Positive Jacobi polynomial sums II, *Amer. J. Math.* **98**(1976), pp. 709–737.

[8] R. Askey and G. Gasper, Inequalities for polynomials, in *The Bieberbach Conjecture, Proceedings of the Symposium on the Occasion of the Proof, Surveys and Monographs* **21** (Amer. Math. Soc., Providence, 1986), pp. 7–32.

[9] L. de Branges, A proof of the Bieberbach conjecture, *Acta Math.* **154**(1985), pp. 137–152.

[10] G. Gasper, Positivity and special functions, in *Theory and Applications of Special Functions,* R. Askey, ed. (Academic Press, New York, 1975), pp. 375–433.

[11] G. Gasper, Positive integrals of Bessel functions, *SIAM J. Math. Anal* **6**(1975), pp. 868–881.

[12] G. Gasper, Positive sums of the classical orthogonal polynomials, *SIAM J. Math. Anal.* **8**(1977), pp. 423–447.

[13] G. Gasper, A short proof of an inequality used by de Branges in his proof of the Bieberbach, Robertson and Milin conjectures, *Complex Variables: Theory Appl.* **7**(1986), pp. 45–50.

[14] G. Gasper, $q$-Extensions of Clausen's formula and of the inequalities used by de Branges in his proof of the Bieberbach, Robertson, and Milin conjectures, *SIAM J. Math. Anal.* **20**(1989), pp. 1019–1034.

[15] G. Gasper, Using symbolic computer algebraic systems to derive formulas involving orthogonal polynomials and other special functions, in *Orthogonal Polynomials: Theory and Practice,* ed. by P. Nevai (Kluwer Academic Publishers, Boston, 1989) pp. 163–179.

[16] G. Gasper and M. Rahman, *Basic Hypergeometric Series,* 2nd edn., *Encyclopedia of Mathematics And Its Applications* **96** (Cambridge University Press, Cambridge, 2004).

[17] J. L. W. V. Jensen, Recherches sur la théorie des équations, *Acta Math.* **36**(1913), pp. 181–195.

[18] R. P. Boas, *Entire Functions* (Academic Press, Inc., New York, 1954).

[19] G. Pólya, Über die algebraisch-funktionentheoretischen Untersuchungen von J. L. W. V. Jensen, *Kgl. Danske Videnskabernes Selskab. Math.-Fys. Medd.* **7(17)** (1927), pp. 3–33; reprinted in his *Collected Papers,* Vol. II, pp. 278–308.

[20] R. Askey and G. Gasper, Solution to Problem 2, *SIAM Activity Group on Orthogonal Polynomials and Special Functions Newsletter* **8(1)**(1997), pp. 18–19; available at http://www.mathematik.uni-kassel.de/~koepf/Siamnews/8-1.pdf.

[21] G. E. Andrews, R. Askey and R. Roy, *Special Functions, Encyclopedia of Mathematics and Its Applications* **71** (Cambridge University Press, Cambridge, 1999).

[22] D. A. Hejhal, On a result of G. Pólya concerning the Riemann $\xi$-function, *J. d'Analyse Math.* **55**(1990), pp. 59–95.

# FUNCTIONS WHOSE MOMENTS FORM A GEOMETRIC PROGRESSION

MOURAD E. H. ISMAIL

*Department of Mathematics, University of Central Florida*
*Orlando, FL 32816, USA*
*E-mail: ismail@math.ucf.edu*

XIN LI

*Department of Mathematics, University of Central Florida*
*Orlando, FL 32816, USA*
*E-mail: xli@math.ucf.edu*

We start with a measure space $L^2[\mathbb{R}, d\mu]$ and give a lower bound for the norm of functions in this space whose first $N$ moments form a geometric progression. Several consequences are investigated including a new criterion for the determinacy of the moment problem. The corresponding questions on the unit circle are also investigated. In particular we give a lower bound for the $L^2$ norm of interpolatory functions in the disk algebra.

*Keywords*: Moments, orthogonal polynomials on $\mathbb{R}$, orthogonal polynomials on the unit circle, moment problems

**Dedicated to Professor Dan Waterman**

## 1. Introduction

Assume that $\mu$ is a positive measure on the real line $\mathbb{R}$ such that the moments $\int_{\mathbb{R}} x^n d\mu$ exist for all $n, n = 0, 1, \cdots$ and we normalize $\mu$ by requiring that $\int_{\mathbb{R}} d\mu = 1$. Let $\{p_n(x)\}$ be the corresponding orthonormal polynomials with positive leading terms. They satisfy a three term recurrence relation [1], [5],

$$xp_n(x) = a_{n+1}p_{n+1}(x) + b_n p_n(x) + a_n p_{n-1}(x), \tag{1}$$

and there is no loss of generality in assuming $p_0(x) = 1, p_1(x) = (x - b_0)/a_1$. Moreover, for all $n \geq 0$, $b_n \in \mathbb{R}$ and $a_{n+1} > 0$. Furthermore we have the Christoffel-Darboux identities [1], [5]

$$\sum_{k=0}^{N} p_k(x)p_k(y) = a_{N+1}\frac{p_{N+1}(x)p_N(y) - p_N(x)p_{N+1}(y)}{x-y}, \tag{2}$$

$$\sum_{k=0}^{N} [p_k(x)]^2 = a_{N+1}[p_N(x)p'_{N+1}(x) - p'_N(x)p_{N+1}(x)]. \tag{3}$$

If the support of $\mu$ is contained in the interval $[0, 2\pi]$, let $\{\varphi_k(z)\}$ be the orthonormal polynomials with respect to $\mu$ on the unit circle, that is,

$$\int_0^{2\pi} \varphi_m(z)\overline{\varphi_n(z)}d\mu(\theta) = \delta_{m,n}, \quad z = e^{i\theta}, \quad m, n = 0, 1, 2, .... \tag{4}$$

The purpose of this note is to prove Theorems 1.1–1.2, and 1.4–1.6; and their corollaries.

**Theorem 1.1.** *Let $\mu$ and $\{p_n(x)\}$ be as above and that $f \in L^2[\mathbb{R}, d\mu]$. If*

$$\int_{\mathbb{R}} x^j f(x)d\mu(x) = c^k, k = 0, 1, \cdots, N,$$

*then we have*

$$\int_{\mathbb{R}} |f(x)|^2 d\mu(x) \geq \sum_{k=0}^{N} |p_k(c)|^2 = \begin{cases} a_{N+1}[p_N(c)p'_{N+1}(c) - p'_N(c)p_{N+1}(c)] & \text{if } c \in \mathbb{R} \\ \frac{a_{N+1}}{c-\overline{c}}[p_{N+1}(c)p_N(\overline{c}) - p_N(c)p_{N+1}(\overline{c})] & \text{if } c \notin \mathbb{R}. \end{cases} \tag{5}$$

*Moreover "=" in* (5) *is achieved if and only if*

$$f(x) = \sum_{k=0}^{N} p_k(c)p_k(x). \tag{6}$$

Theorem 1.1 should be contrasted with the familiar property: There is no function $f$ in $L[[0,1], dx]$ such that $f \geq 0, \int_0^1 f(x)dx = 1, \int_0^1 xf(x)dx = \alpha, \int_0^1 x^2 f(x)dx = \alpha^2$ for some $\alpha > 0$.

Recall that the moments $\int_{\mathbb{R}} x^n d\mu(x)$ may or may not determine $\mu$ uniquely. The moment problem is called determinate [1] if the moments determine a unique positive measure $\mu$. A moment problem is called indeterminate if it is not determinate. Theorem 1.1 can be used to establish the following new criterion for the determinacy of the moment problem.

**Theorem 1.2.** *Let $c \in \mathbb{C} \setminus \mathbb{R}$ and $\mu$ be a probability measure on $\mathbb{R}$ with moments of all orders. Then the moment problem associated with $\mu$ is indeterminate if and only if there is a function $f \in L^2[\mathbb{R}, \mu]$ such that*

$$\int_{\mathbb{R}} f(x)x^j d\mu(x) = c^j, \; j = 0, 1, .... \tag{7}$$

**Theorem 1.3.** *Let $f \in L^2[[0,1], (1-x)^\alpha x^\beta dx]$. If*

$$\int_0^1 x^j f(x) \frac{\Gamma(\alpha+\beta+2)}{\Gamma(\alpha+1)\Gamma(\beta+1)} (1-x)^\alpha x^\beta dx = 1, \quad j = 0, 1, \cdots, N-1,$$

*then*

$$\int_0^1 |f(x)|^2(1-x)^\alpha x^\beta dx \geq \frac{(\alpha+2)_{N-1}(\alpha+\beta+2)_{N-1}}{(N-1)!\,(\beta+1)_{N-1}}. \tag{8}$$

*Moreover, " = " is attained if and only if*

$$f(x) = \sum_{k=0}^{N-1} \frac{(\alpha+1)_k(\alpha+\beta+1)_k(\alpha+\beta+2k+1)}{k!(\beta+1)_k(\alpha+\beta+1)} \times {}_2F_1\left(\begin{matrix} -k, k+\alpha+\beta+1 \\ \alpha+1 \end{matrix}\middle| \frac{1-x}{2}\right). \tag{9}$$

The special case $\alpha = \beta = 0$ of Theorem 1.3 is the following:

**Corollary 1.1.** *Assume that* $f \in L^2[[0,1], dx]$. *If* $\int_0^1 x^j f(x)dx = 1, j = 0, 1, \cdots, N-1$, *then*

$$\int_0^1 |f(x)|^2 dx \geq N^2. \tag{10}$$

Corollary 1.1 with the additional condition that $f$ is continuous is a Monthly problem, see [3].

We now come to orthogonal polynomials on the unit circle. Assume $\mu$ is a positive measure on $[0, 2\pi]$ and $\int_0^{2\pi} d\mu(\theta) = 1$. The corresponding orthonormal polynomials will be denoted by $\{\varphi(z)\}$, $z = e^{i\theta}$. All results stated so far can be extended to the unit circle case. We give one such example to demonstrate this in the next theorem.

**Theorem 1.4.** *Let* $f \in L^2[[0, 2\pi], d\mu]$. *If*

$$\int_0^{2\pi} e^{-ik\theta} f(\theta) d\mu(\theta) = e^{-ik\alpha}$$

*for* $k = 0, 1, ..., N$, *then we have*

$$\int_0^{2\pi} |f(\theta)|^2 d\mu(\theta) \geq \sum_{k=0}^{N} |\varphi_k(e^{i\alpha})|^2. \tag{11}$$

We have considered either the real line $\mathbb{R}$ or the unit circle in the complex plane $\mathbb{C}$. Next, we formulate a more general result. Let $G$ be a Borel set in $\mathbb{C}$ and let $\mu$ be a finite measure on $G$. Consider the function space $L^2[G, d\mu]$ of functions square integrable on $G$. Define the usual inner product by

$$\langle f, g\rangle = \int_G f(z)\overline{g(z)} d\mu \quad \text{for } f, g \in L^2[G, d\mu].$$

Let $\{x_k\}_{k=0}^n$ be a linear independent set of function from $L^2[G, d\mu]$ and let $P_n = \text{span}(\{x_k\}_{k=0}^n)$. Let $\{p_k\}_{k=0}^n$ be an orthonormal basis in $P_n$.

**Theorem 1.5.** *Assume* $f \in L^2[G, d\mu]$ *and*

$$\int_G \overline{x_k(z)} f(z) d\mu = b_k, \quad k = 0, 1, ..., N.$$

*Then*

$$\int_G |f(z)|^2 d\mu \geq \mathbf{b}^T A^{-T} A^{-1} \mathbf{b},$$

*where* $\mathbf{b}^T = (b_0, b_1, ..., b_n)$ *and* $A = (\overline{a_{jk}})_{j,k=0}^n$ *satisfying*

$$x_j(z) = \sum_{k=0}^n a_{jk} p_k(z), \quad j = 0, 1, ..., N.$$

Section 2 contains proofs of the results stated so far and some of their corollaries. It also contains variations on the theorems stated so far.

## 2. Proofs

**Proof of Theorem 1.1.** Let $\sum_{n=0}^\infty c_n p_n(x)$ be the orthogonal series of $f(x)$ and assume that $p_n(x) = \sum_{k=0}^n a_{n,k} x^k$. Then

$$c_m = \int_{\mathbb{R}} f(x) p_m(x) d\mu(x) = \sum_{k=0}^n a_{m,k} \int_{\mathbb{R}} f(x) x^k d\mu(x) = p_m(c).$$

The Bessel inequality implies

$$\int_{\mathbb{R}} |f(x)|^2 d\mu(x) \geq \sum_{k=0}^\infty |c_k|^2 \geq \sum_{k=0}^N |p_k(c)|^2.$$

Now the "$\geq$" step in (5) follows. The "$=$" step follows from the Christoffel-Darboux formula (2)–(3). If "$\geq$" in (5) is "$=$" then the coefficients $c_j$ must vanish for all $j > N$, hence $f$ is the polynomial given by (6). □

Observe that Theorem 1.1 continues to hold if we only assume that the moments $\int_{\mathbb{R}} x^m d\mu(x)$ exist, for $m = 0, \ldots, 2N-1$, and $\mu$ is a probability measure.

**Proof of Theorem 1.2.** Let $f \in L^2[\mathbb{R}, d\mu]$. If the condition (7) holds we apply Theorem 1.1, let $N \to \infty$ and conclude that $\sum_{n=0}^\infty |p_n(c)|^2$ converges. Hence the moment problem is indeterminate [1]. Conversely if the moment problem is indeterminate then $\{p_n(c)\} \in \ell^2$ and the function $f$ whose orthogonal series is $\sum_{n=0}^\infty p_n(c) p_n(x)$ is in $L^2[\mathbb{R}, d\mu]$. Write $x^m = \sum_{k=0}^m \lambda_{m,k} p_k(x)$. The function $x^m f(x)$ is in $L^1[\mathbb{R}, d\mu]$, and

$$\int_{\mathbb{R}} x^m f(x) d\mu(x) = \sum_{k=0}^m \lambda_{m,k} \int_{\mathbb{R}} f(x) p_k(x) d\mu(x) = \sum_{k=0}^m \lambda_{m,k} p_k(c) = c^m,$$

holds for all nonnegative integers $m$, and the proof is complete. □

**Proof of Theorem 1.3.** In this case the orthonormal polynomials are

$$p_n^{(\alpha,\beta)}(x) = A(\alpha,\beta,n)\ {}_2F_1\left(\left.\begin{matrix}-n, n+\alpha+\beta+1\\ \alpha+1\end{matrix}\right| 1-x\right), \tag{12}$$

and

$$\frac{d}{dx}p_n^{(\alpha,\beta)}(x) = \frac{n(n+\alpha+\beta+1)}{(\alpha+1)}A(\alpha,\beta,n)\ {}_2F_1\left(\left.\begin{matrix}1-n, n+\alpha+\beta+2\\ \alpha+2\end{matrix}\right| 1-x\right) \tag{13}$$

where

$$A(\alpha,\beta,n) = \sqrt{\frac{(\alpha+1)_n(\alpha+\beta+1)_n(\alpha+\beta+2n+1)}{n!\,(\beta+1)_n\,(\alpha+\beta+1)}}.$$

Moreover (4.2.9) in [5] yields

$$a_{n+1}^2 = \frac{(n+1)(\alpha+n+1)(\beta+n+1)(\alpha+\beta+n+1)}{(\alpha+\beta+2n)(\alpha+\beta+2n+2)^2(\alpha+\beta+2n+3)}.$$

The result now follows from Theorem 1.1 and the fact that $p_n^{(\alpha,\beta)}(1) = A(\alpha,\beta,n)$. □

Corollary 1.1 is the case $\alpha = \beta = 0$ of Theorem 1.3, so the $p_n$'s are the Legendre polynomials renormalized to become orthonormal on $[0,1]$. Thus $p_n(x) = \sqrt{2n+1}\,{}_2F_1(-n, n+1; 1; 1-x)$ and $p_n(1) = \sqrt{2n+1}$, [2], [5]. Thus $\sum_{k=0}^{N-1}[p_k(1)]^2 = N^2$.

The proof of Theorem 1.4 is similar to the proof of Theorem 1.1 and will be omitted.

**Corollary 2.1.** *Assume that* $\int_0^{2\pi}|f(\theta)|^2 d\theta < \infty$. *If*

$$\int_0^{2\pi} e^{-ik\theta} f(\theta)d\theta = 1$$

*for* $k = 0, 1, ..., N-1$, *then*

$$\int_0^{2\pi} |f(\theta)|^2 d\theta \geq \frac{N}{2\pi}$$

*holds.*

**Proof.** The orthonormal polynomials with respect to $d\theta$ on the unit circle are $\{z^n/\sqrt{2\pi}\}$ with $z = e^{i\theta}$ for $k = 0, 1, ....$ Take $\alpha = 0$ in Theorem 1.4 to obtain the result. □

**Corollary 2.2.** *Assume that* $\int_0^{2\pi}|f(\theta)|^2 d\theta < \infty$ *and* $|\zeta| < 1$. *If*

$$\int_0^{2\pi} e^{-ik\theta} f(\theta)\frac{1-|\zeta|^2}{|z-\zeta|^2}\frac{d\theta}{2\pi} = e^{-i\alpha}, \quad z = e^{ik\theta},$$

*for* $k = 0, 1, ..., n$, *then we have*

$$\int_0^{2\pi} |f(\theta)|^2 \frac{1-|\zeta|^2}{|z-\zeta|^2}\frac{d\theta}{2\pi} \geq 1 + \frac{n|e^{i\alpha}-\zeta|^2}{1-|\zeta|^2}.$$

**Proof.** As a special case of the Bernstein weight, the Poisson weight

$$\frac{1-|\zeta|^2}{|z-\zeta|^2}\frac{d\theta}{2\pi}$$

with $z = e^{i\theta}$ on the unit circle has orthonormal polynomials

$$1,\ (z-\zeta)/\sqrt{1-|\zeta|^2},\ z(z-\zeta)/\sqrt{1-|\zeta|^2}, ..., z^{k-1}(z-\zeta)/\sqrt{1-|\zeta|^2}$$

for $k = 1, 2, ....$ □

**Proof of Theorem 1.5.** Let $g$ denote the projection of $f$ onto $P_n$. Then

$$b_j = \langle f, x_j\rangle = \langle g, x_j\rangle, \quad j = 0, 1, ..., n.$$

Write $g(z) = \sum_{k=0}^n g_k p_k(z)$. So, the $g_k$'s must satisfy

$$b_j = \sum_{k=0}^n g_k \langle p_k, x_j\rangle, \quad j = 0, 1, ..., n.$$

But note that

$$\langle p_k, x_j\rangle = \langle p_k, \sum_{l=0}^n a_{jl} p_l\rangle = \sum_{l=0}^n \overline{a_{jl}}\langle p_k, p_l\rangle = \overline{a_{jk}}, \quad j, k = 0, 1, ..., n.$$

So,

$$b_j = \sum_{k=0}^n g_k \overline{a_{jk}}, \quad j = 0, 1, ..., n.$$

Thus $\mathbf{b} = A\mathbf{g}$ if we write $\mathbf{g} = (g_0, g_1, ..., g_n)^T$. Therefore, $\mathbf{g} = A^{-1}\mathbf{b}$. Finally, we have

$$\int_G |f(z)|^2 d\mu = \langle f, f\rangle \geq \langle g, g\rangle = \|\mathbf{g}\|^2 = \mathbf{g}^T\mathbf{g} = \mathbf{b}^T A^{-T} A^{-1}\mathbf{b},$$

and the proof is complete. □

**Corollary 2.3.** *Under the same assumptions as in Theorem 1.5 and if* $b_k = \overline{x_k(\zeta)}$ *for some* $\zeta \in G$ *and for all* $k = 0, 1, ..., n$, *then*

$$\int_G |f(z)|^2 d\mu \geq \sum_{k=0}^n |p_k(\zeta)|^2.$$

**Proof.** Note that in this case

$$A^{-1}\mathbf{b} = A^{-1}\overline{(x_0(\zeta), x_1(\zeta), ..., x_n(\zeta))^T} = \overline{(p_0(\zeta), p_1(\zeta), ..., p_n(\zeta))^T},$$

by the definition of $A$. Therefore,

$$\mathbf{b}^T A^{-T} A^{-1}\mathbf{b} = (p_0(\zeta), p_1(\zeta), ..., p_n(\zeta))\overline{(p_0(\zeta), p_1(\zeta), ..., p_n(\zeta))^T} = \sum_{k=0}^n |p_k(\zeta)|^2.$$ □

Our next result is an interesting surprise on functions from the disk algebra that satisfy certain interpolatory assumptions.

**Corollary 2.4.** *Assume that $f(z)$ is analytic for $|z|<1$ and continuous for $|z|\leq 1$ and assume that $|\zeta_k|<1$, $k=0,1,...,n$, are $n+1$ distinct points. If*

$$f(\zeta_k)=\frac{1}{1-\zeta_k e^{-i\alpha}}$$

*for $k=0,1,...,n$, then we have*

$$\frac{1}{2\pi}\int_0^{2\pi}|f(e^{i\theta})|^2 d\theta \geq \sum_{k=0}^{n}\frac{1-|\zeta_k|^2}{|e^{i\alpha}-\zeta_k|^2}.$$

**Proof.** By the Cauchy integral formula, we have

$$f(\zeta_k)=\frac{1}{2\pi i}\int_{|z|=1}\frac{f(z)}{z-\zeta_k}dz,$$

which can be written as

$$\frac{1}{2\pi}\int_{|z|=1}\frac{f(z)z\,dz}{z-\zeta_k\ \ zi}=\frac{1}{2\pi}\int_0^{2\pi}\frac{f(z)}{1-\zeta_k\overline{z}}d\theta=\frac{1}{2\pi}\int_0^{2\pi}\overline{\left(\frac{1}{1-\overline{\zeta_k}z}\right)}f(z)d\theta.$$

Thus, the interpolatory assumptions on $f$ can be written as

$$\frac{1}{2\pi}\int_0^{2\pi}\overline{\left(\frac{1}{1-\overline{\zeta_k}z}\right)}f(z)d\theta=\frac{1}{1-\zeta_k e^{-i\alpha}},\quad \text{for } k=0,1,...,n. \tag{14}$$

Now, recall the Takenaka-Malmquist system [7, p.224]:

$$\psi_k(z)=\frac{\sqrt{1-|\zeta_k|^2}}{1-\overline{\zeta_k}z}\prod_{j=0}^{k-1}\frac{z-\zeta_j}{1-\overline{\zeta_j}z},\quad k=0,1,...,n.$$

It is known that this system is orthonormal with respect to $d\theta/(2\pi)$ on the unit circle. So, $\{\psi_k\}_{k=0}^n$ forms an orthonormal basis for the space $R_{n,n+1}$ of rational functions of degree [a] at most $(n,n+1)$ with poles among the points $1/\overline{\zeta_k}$, $k=0,1,...,n$. Now, using (14), by Corollary 2.3, we have

$$\frac{1}{2\pi}\int_0^{2\pi}|f(z)|^2 d\theta \geq \sum_{k=0}^{n}|\psi_k(e^{i\theta})|^2,$$

which is equivalent to the desired result. □

[a] The degree of a rational function is the ordered pair consisting of the numerator degree and denominator degree.

## References

[1] N. I. Akhiezer, *The Classical Moment Problem and Some Related Questions in Analysis*, English translation, Oliver and Boyed, Edinburgh, 1965.

[2] G. E. Andrews, R. A. Askey, and R. Roy, *Special Functions*, Cambridge University Press, Cambridge, 1999.

[3] P. P. Dályay, Problem 11248. American Math. Monthly, **113** (2006), 760.

[4] G. Gasper and M. Rahman, *Basic Hypergeometric Series*, Second Edition, Cambridge University Press, Cambridge, 2004.

[5] M. E. H. Ismail, *Classical and Quantum Orthogonal Polynomials in one Variable*, Cambridge University Press, Cambridge, 2005.

[6] G. Szegö, *Orthogonal Polynomials*, Fourth Edition, Amer. Math. Soc., Providence, RI, 1975.

[7] J. L. Walsh, *Interpolation and Approximation*, Amer. Math. Soc., Providence, RI, 1960.

# CHARACTERIZATION OF SCALING FUNCTIONS IN A FRAME MULTIRESOLUTION ANALYSIS IN $H^2_G$

K. S. KAZARIAN

*Departamento de Matemáticas, Universidad Autónoma de Madrid*
*Madrid, 28049, Spain*
*E-mail: kazaros.kazarian@uam.es*

A. SAN ANTOLÍN

*Departamento de Matemáticas, Universidad Autónoma de Madrid*
*Madrid, 28049, Spain*
*E-mail: angel.sanantolin@uam.es*

This is partly a survey article. We present a survey of results related with the characterization of scaling functions of multiresolution analyses. Given a linear invertible map $A : \mathbb{R}^n \to \mathbb{R}^n$ such that $A(\mathbb{Z}^n) \subset \mathbb{Z}^n$ and all (complex) eigenvalues of $A$ have absolute value greater than one we give a characterization of scaling functions of a frame multiresolution analysis on some closed subspaces of $L^2(\mathbb{R}^n)$ denoted by $H^2_G$. As a corollary we obtain that if the density number of the set $G$, $|G|_n > 0$ at the origin is less than one then no any function $g \in L^1(\mathbb{R}^n)$ can be a scaling function of an $H^2_G$-FMRA associated with $A$. Putting together results obtained in [22], [12] one observes that for any measurable $A^*$-invariant set $G$ of positive measure always exists some $H^2_G$-MRA associated with $A$.

*Keywords*: Multiresolution analysis; scaling function; Fourier transform; approximate continuity.

**Dedicated to Professor Daniel Waterman**

## 1. Introduction

A multiresolution analysis (MRA) is a general method introduced by Mallat [31] and Meyer [32] for constructing wavelets. By an MRA on $\mathbb{R}^n (n \geq 1)$, one means a sequence of closed subspaces $V_j$, $j \in \mathbb{Z}$ of the Hilbert space $L^2(\mathbb{R}^n)$ that satisfies the following conditions:

(i) $\forall j \in \mathbb{Z}, \quad V_j \subset V_{j+1}$;
(ii) $\forall j \in \mathbb{Z}, \quad f(\mathbf{x}) \in V_j \Leftrightarrow f(2\mathbf{x}) \in V_{j+1}$;
(iii) $W = \overline{\cup_{j \in \mathbb{Z}} V_j} = L^2(\mathbb{R}^n)$;
(iv) There exists a function $\phi \in V_0$, that is called *scaling function*, such that the system $\{\phi(\mathbf{x} - \mathbf{k})\}_{\mathbf{k} \in \mathbb{Z}^n}$ is an orthonormal basis for $V_0$.

We will consider MRA in a general context, where instead of the dyadic dilation one has a dilation given by a fixed linear invertible map $A : \mathbb{R}^n \to \mathbb{R}^n$ such that

(*) $A(\mathbb{Z}^n) \subset \mathbb{Z}^n$ and all (complex) eigenvalues of $A$ have absolute value greater than one.

Given such a linear map $A$ one defines an $A-$MRA as a sequence of subspaces $V_j$, $j \in \mathbb{Z}$ of the Hilbert space $L^2(\mathbb{R}^n)$ (see [30], [21], [38], [39]) that satisfies the conditions (i), (iii), (iv) and

(ii$_1$) $\forall j \in \mathbb{Z}, \;\; f(\mathbf{x}) \in V_j \Leftrightarrow f(A\mathbf{x}) \in V_{j+1}$.

Many authors have studied conditions under which a given function can be a scaling function for an MRA. The Fourier transform plays an important role in the study of the above problem, which can be explained by the presence of the operations of translation and dilation in the definition of an MRA.

In this paper we adopt the convention that the Fourier transform of a function $f \in L^1(\mathbb{R}^n) \cap L^2(\mathbb{R}^n)$ is defined by

$$\widehat{f}(\mathbf{y}) = \int_{\mathbb{R}^n} f(\mathbf{x}) e^{-2\pi i \mathbf{x}\cdot\mathbf{y}} d\mathbf{x}.$$

Suppose that $\phi$ is a given function in $L^2(\mathbb{R})$ such that $\{\phi(\cdot - k)\}_{k\in\mathbb{Z}}$ is an orthonormal system in $L^2(\mathbb{R})$. Let $V_0$ be the closure of the finite linear combinations of that system. Moreover, let us suppose that conditions (i) and (ii) are satisfied. The aim of this paper is to discuss results related with the following

Problem. What additional condition one needs to guarantee that (iii) holds?

This question was studied by several authors (cf. [31], [27], [8], [17], [4], [25], [39], [29], [26], [11] and others). We will discuss known results bringing them in the chronological order.

The following theorem should be attributed to S. G. Mallat [31].

**Theorem A.** *Let $V_j$, $j \in \mathbb{Z}$ be a sequence of closed subspaces of $L^2(\mathbb{R})$ satisfying* (i), (ii) *and* (iv) *with $n = 1$, where $\widehat{\phi}$ is continuously differentiable and satisfies the following conditions:*

$$\exists C > 0, \forall y \in \mathbb{R}, \quad | \widehat{\phi}(y) | \leq C(1+y^2)^{-1} \text{ and } | \frac{d}{dy}\widehat{\phi}(y) | \leq C(1+y^2)^{-1}.$$

*Then the condition* (iii) *holds if and only if $|\widehat{\phi}(0)| = 1$.*

Afterwards a closely related problem was studied by R. Q. Jia and C. A. Micchelli [27].

**Theorem B.** *Let $\phi$ be a function defined on $\mathbb{R}^n$ such that*

$$\int_{[0,1)^n} |(\sum_{\mathbf{k}\in\mathbb{Z}^n} |\phi(\mathbf{x}-\mathbf{k})|)|^p d\mathbf{x} < \infty, \qquad (1 \leq p < \infty),$$

*and $\sum_{\mathbf{k}\in\mathbb{Z}^n} \phi(\mathbf{x}-\mathbf{k}) = 1$. Then, for any $f \in L^p(\mathbb{R}^n)$,*

$$\| f - \sum_{\mathbf{k}\in\mathbb{Z}^n} a_h(\mathbf{k})\phi(h^{-1}\cdot - \mathbf{k}) \|_p \to 0 \qquad as \quad h \to 0,$$

*where*

$$a_h(\mathbf{k}) = a_h(f, \mathbf{k}) := h^{-n} \int_{h\mathbf{k}+[0,h)^n} f(\mathbf{x}) d\mathbf{x} = \int_{[0,h)^n} f(h(\mathbf{x}+\mathbf{k})) d\mathbf{x}.$$

The interested reader can find references related with approximations by linear combinations of integer (multi-integer) translates of a function in [27, Remark 2.2].

To the best of our knowledge the following proposition appeared first in the monograph of I. Daubechies [17].

**Theorem C.** *Let $V_j, j \in \mathbb{Z}$, be a sequence of closed subspaces of $L^2(\mathbb{R})$ satisfying (ii) and (iv) with $n = 1$, where $\widehat{\phi}$ is such that $\widehat{\phi}$ is a bounded function, continuous near $y = 0$, and such that, moreover, $\widehat{\phi}(0) \neq 0$. Then $\overline{\cup_{j\in\mathbb{Z}} V_j} = L^2(\mathbb{R}^n)$.*

The following observations were made in [17] related with Theorem C:

• the hypothesis about the continuity of $\widehat{\phi}$ is not necessary;

• if the condition (i) holds and $\widehat{\phi}$ is bounded and continuous at 0, then $\widehat{\phi}(0) \neq 0$ is a necessary condition;

• the argument in the proof gives that $|\widehat{\phi}(0)| = 1$.

Somewhat later, P. Wojtaszczyk [39] observed that in Theorem C the boundedness of $\widehat{\phi}$ for almost all $y$ is redundant.

Historically, the first complete descriptions of conditions under which $\cup_{j\in\mathbb{Z}} V_j$ is dense in $L^2(\mathbb{R}^n)$ were given by W. R. Madych [30] and by C. de Boor, R. DeVore and A. Ron [4]. The following result for an $A$-MRA was proved by W. R. Madych.

**Theorem D.** *Let $\{V_j\}_{j\in\mathbb{Z}}$ be a sequence of closed subspaces of $L^2(\mathbb{R}^n)$ satisfying ($ii_1$) and (iv). Let $P_j f$ denote the orthogonal projection of $f$ onto $V_j$. Then, the following conditions are equivalent:*

- *For all $f$ in $L^2(\mathbb{R}^n)$*

$$\lim_{j\to\infty} \| f - P_j f \| = 0.$$

- *The function $\phi$ in (iv) satisfies*

$$\lim_{j\to\infty} \frac{1}{|(A^*)^{-j} Q|} \int_{(A^*)^{-j} Q} |\widehat{\phi}(\mathbf{t})|^2 d\mathbf{t} = 0$$

*for every cube $Q$ of finite diameter in $\mathbb{R}^n$, where $A^*$ is the adjoint of $A$.*

The result proved by C. de Boor, R. DeVore and A. Ron is the following.

**Theorem E.** *Let $\phi \in L^2(\mathbb{R}^n)$, and let $V_0$ be the $L^2(\mathbb{R}^n)$-closure of the finite linear combinations of the multi-integer translates of $\phi$, moreover, for any $j \in \mathbb{Z}$, let*

$$V_j = \{f(2^j \mathbf{x}) \ : \ f \in V_0\}.$$

*Suppose, moreover, that the sequence of subspaces $V_j$, $j \in \mathbb{Z}$, satisfies (i) and (ii). Then, the condition (iii) holds if and only if*

$$\cup_{j\in\mathbb{Z}} \, (2^j \operatorname{supp} \widehat{\phi}) = \mathbb{R}^n \qquad \text{(modulo a null-set)},$$

*where*

$$\operatorname{supp} \widehat{\phi} = \{\mathbf{t} \in \mathbb{R}^n \ : \ \widehat{\phi}(\mathbf{t}) \neq 0\}.$$

The following result is an immediate consequence of Theorem E (cf. [1]).

**Theorem F.** *Let $\phi \in L^2(\mathbb{R}^n)$ and set $V_0 = \overline{span}\{\phi(\cdot + \mathbf{k}) \ : \ \mathbf{k} \in \mathbb{Z}^n\}$ and $V_j = \{f(2^j\cdot) \ : \ f \in V_0\}$ for each $j \in \mathbb{Z}^n$. Suppose $V_0 \subset V_1$. If $|\widehat{\phi}| > 0$ almost everywhere on some neighborhood of 0, then $\overline{\cup_{j\in\mathbb{Z}} V_j} = L^2(\mathbb{R}^n)$.*

The following improvement of Theorem C appears in the book of E. Hernández and G. Weiss [25].

**Theorem G.** *Let $V_j$, $j \in \mathbb{Z}$, be a sequence of closed subspaces of $L^2(\mathbb{R})$ satisfying* (i), (ii) *and* (iv) *with $n = 1$, where $\widehat{\phi}$ is such that $|\ \widehat{\phi}\ |$ is continuous at the origin. Then the condition (iii) holds if and only if $\widehat{\phi}(0) \neq 0$.*

Somewhat later the problem concerning the characterization of scaling functions of an MRA was formulated in a paper by R. Strichartz [38]. Evidently he was not acquainted with [30] and [4].

Apparently Strichartz's question motivated E. Hernández, X. Wang and G. Weiss [26] (see also [25, Theorem 5.2, p. 382]), to prove the following result.

**Theorem H.** *Let $V_j$, $j \in \mathbb{Z}$, be a sequence of closed subspaces of $L^2(\mathbb{R})$ satisfying* (i), (ii) *and* (iv) *with $n = 1$. Then, the condition* (iii) *holds if and only if*

$$\lim_{j\to\infty} |\widehat{\phi}(2^{-j}y)| = 1 \quad \textit{for a.e. } y \in \mathbb{R}.$$

Another set of necessary and sufficient conditions were obtained by R. A. Lorentz, W. R. Madych and A. Sahakian [29].

**Theorem I.** *Let $V_j$, $j \in \mathbb{Z}$, be a sequence of closed subspaces of $L^2(\mathbb{R})$ satisfying* (i), (ii) *and* (iv) *with $n = 1$. Then the condition* (iii) *is equivalent to the following conditions*:

- $\lim_{j\to\infty} |\widehat{\phi}(2^{-j}y)|$ *exists and is positive for a.e. $y \in \mathbb{R}$;*
- *The set $\{y \in \mathbb{R} \ : \ |\ \widehat{\phi}(y)\ | > 0\}$ is dyadically absorbing, i.e., for a.e. $y \in \mathbb{R}$, there exists a positive integer $j_0$, which may depend on $y$, such that if $j \geq j_0$ then $|\ \widehat{\phi}(2^{-j}y)\ | > 0$.*
- $\lim_{j\to\infty} 2^j \phi * \widetilde{\phi}(2^j y)$ *exists in the distributional sense and is a nonzero multiple of the Dirac distribution at the origin. Here $\widetilde{\phi}(y) = \overline{\phi(-y)}$ and $*$ denotes the usual convolution.*

A generalization of Theorem I for an $A$-MRA generated by several scaling functions was formulated by A. Calogero [7].

A characterization of scaling functions of an MRA in probabilistic terms was given by V. Dobrić, R. F. Gundy and P. Hitczenko [19].

A somewhat deeper understanding of the relation between the behavior of the function $\widehat{\phi}$ in the neighborhood of the origin and the condition (iii) is achieved in the paper [11] by P. Cifuentes, K. S. Kazarian and A. San Antolín. We prefer to speak about the results obtained in [11] in Section 3 where a characterization of scaling functions for $A$-MRA's defined on some subspaces of $L^2(\mathbb{R}^n)$ will be obtained. We will consider also similar characterization in a more general context of $A$-FMRA's for some spaces $H^2_G$. The spaces $H^2_G$ can be considered as generalization of the classical Hardy spaces on the real line. In the literature there are several other generalizations of MRA's which are beyond the scope of the present paper and we will not discuss them.

## 2. Spaces $H^2_G$

From now on we shall use the same notation for the linear map and its matrix with respect to the canonical base. For a given linear invertible map $\mathcal{A} : \mathbb{R}^n \to \mathbb{R}^n$ consider the unitary operator $D_{\mathcal{A}} : L^2(\mathbb{R}^n) \to L^2(\mathbb{R}^n)$ defined by

$$(D_{\mathcal{A}}f)(\mathbf{t}) = d_{\mathcal{A}}^{\frac{1}{2}} f(\mathcal{A}\mathbf{t}), \qquad \text{for any} \quad f \in L^2(\mathbb{R}^n),$$

where $d_{\mathcal{A}} = |\det \mathcal{A}|$. It can be easily verified that

$$D_{\mathcal{A}}^{-1} = D_{\mathcal{A}^{-1}}. \tag{1}$$

For any $\mathbf{u} \in \mathbb{R}^n$ let $T_{\mathbf{u}}$ be the operator of translation by the vector $\mathbf{u}$ :

$$(T_{\mathbf{u}}f)(\cdot) = f(\cdot - \mathbf{u}), \qquad \text{for any} \quad f \in L^2(\mathbb{R}^n).$$

We leave to the reader to verify the following equality

$$\widehat{T_{\mathbf{k}} D_{\mathcal{A}}^j f} = D_{\mathcal{A}^*}^{-j} E_{\mathbf{k}} \widehat{f}, \qquad \text{where} \quad E_{\mathbf{k}} f(\mathbf{t}) = e^{-2\pi i \mathbf{k}\cdot\mathbf{t}} f(\mathbf{t}). \tag{2}$$

Let $L$ be a subset of a separable Hilbert space $\mathbb{H}$. The vector space generated by all finite linear combinations of elements in $L$ is denoted by span$L$ and its closure by $\overline{\text{span}}L$. The orthogonal complement of $L$ in $\mathbb{H}$ is denoted by $L^\perp$.

We shall say that a linear subspace $\mathbb{E} \subseteq L^2(\mathbb{R}^n)$ is an $\mathcal{A}$-reducing subspace if $D_{\mathcal{A}}\mathbb{E} = \mathbb{E}$ and $T_\ell \mathbb{E} = \mathbb{E}$ for any $\ell \in \mathbb{Z}^n$.

If $\mathcal{A} : \mathbb{R}^n \to \mathbb{R}^n$ is an expansive map (that is if all the eigenvalues of $\mathcal{A}$ have modulus greater than one) then a complete description of the $\mathcal{A}$-reducing subspaces was given by X. Dai, Y. Diao, Q. Gu and D. Han [13]. Observe that if $\mathbb{E} \subseteq L^2(\mathbb{R}^n)$ is an $\mathcal{A}$-reducing subspace for a linear invertible mapping $\mathcal{A}$ then it will be also an $\mathcal{A}^{-1}$-reducing subspace for the inverse mapping $\mathcal{A}^{-1}$.

We will obtain the result in [13] in a slightly easier way using only elementary tools from the Fourier Analysis. The following result is well known to the experts (see [24], p. 145; cf. [30], p. 278).

**Lemma A.** *Let $\mathbb{E}$ be a closed subspace of $L^2(\mathbb{R}^n)$ such that $T_{\mathbf{u}}\mathbb{E} = \mathbb{E}$ for all $\mathbf{u} \in \mathbb{R}^n$. Then $\mathbb{E}$ consists of all functions whose Fourier transforms are supported on some fixed Lebesgue measurable subset of $\mathbb{R}^n$.*

**Proof of Lemma A.** If $\mathbb{E}^\perp = \emptyset$ then evidently the lemma is true. Suppose that $\mathbb{E}^\perp \neq \emptyset$. Then for any $g \in \mathbb{E}^\perp$ and for every $f \in \mathbb{E}$

$$\int_{\mathbb{R}^n} f(\mathbf{x}+\mathbf{t})\overline{g(\mathbf{t})}d\mathbf{t} = 0$$

for all $\mathbf{x} \in \mathbb{R}^n$ and therefore, by Plancherel's identity

$$\int_{\mathbb{R}^n} e^{2\pi i \mathbf{t}\cdot\mathbf{x}} \widehat{f}(\mathbf{t})\overline{\widehat{g}(\mathbf{t})}d\mathbf{t} = 0.$$

This shows that the Fourier transform of $\widehat{f\overline{g}}$ is identically zero, which yields

$$\widehat{f}(\mathbf{t})\overline{\widehat{g}}(\mathbf{t}) = 0 \quad \text{almost everywhere (a.e.) on} \quad \mathbb{R}^n. \tag{3}$$

That is, $\widehat{f}$ and $\widehat{g}$ have disjoint supports for all $f \in \mathbb{E}$ and for all $g \in \mathbb{E}^\perp$. Consider an orthonormal system of functions $\{\psi_k\}_{k=1}^\infty$ defined on $\mathbb{R}^n$ which is a basis of $\mathbb{E}^\perp$. Let $G_k = \{\mathbf{t} \in \mathbb{R}^n : \widehat{\psi}_k(\mathbf{t}) = 0\}$ and define $G = \bigcap_{k=1}^\infty G_k$. Since $\{\psi_k\}_{k=1}^\infty$ is a basis of $\mathbb{E}^\perp$ we get

$$\operatorname{supp} \widehat{g} \subseteq G^c \quad \text{for any} \quad g \in \mathbb{E}^\perp, \tag{4}$$

where $G^c = \mathbb{R}^n \setminus G$. On the other hand from (3) it follows that

$$\operatorname{supp} \widehat{f} \subseteq G \quad \text{for any} \quad f \in \mathbb{E}, \tag{5}$$

and the result follows. □

The following fact is verified easily (e.g. [11], pp. 1020-1021).

**Remark 2.1.** Let $\mathcal{A} : \mathbb{R}^n \to \mathbb{R}^n$ be a linear invertible map such that the set $\bigcup_{j=-\infty}^\infty \mathcal{A}^j(\mathbb{Z}^n)$ is dense in $\mathbb{R}^n$. Then any $\mathcal{A}$-reducing subspace will be invariant with respect to translations.

It is well known that when $\mathcal{A}$ is an expansive map the set $\bigcup_{j=0}^\infty \mathcal{A}^j(\mathbb{Z}^n)$ is dense in $\mathbb{R}^n$ (e.g. [13], p. 3261 or [11], p. 1020). Thus, by Lemma A, there exists a Lebesgue measurable subset $G \subseteq \mathbb{R}^n$ such that the conditions (5) and (4) hold. We now investigate further properties of the set $G$. The condition $D_\mathcal{A}\mathbb{E} = \mathbb{E}$ yields $D_\mathcal{A}\mathbb{E}^\perp = \mathbb{E}^\perp$. Hence, using the well known and easily verifiable relation $\widehat{D_\mathcal{A}} = D_{\mathcal{A}^{*-1}}$ we obtain $\mathcal{A}^{*-1}(G) \subseteq G$ and $\mathcal{A}^{*-1}(G^c) \subseteq G^c$. Combining these observations we see that $\mathcal{A}^*(G) = G$.

**Definition 2.1.** We will say that a set $G \subset \mathbb{R}^n$ is an $\mathcal{A}$–invariant set or an $\mathcal{A}$–set if $\mathcal{A}(G) = G$.

If $\mathcal{A} : \mathbb{R}^n \to \mathbb{R}^n$ is a linear invertible map and $\mathbb{E}$ is a closed subspace of $L^2(\mathbb{R}^n)$ such that both conditions (5), (4) hold for a Lebesgue measurable $\mathcal{A}^*$–set $G \subset \mathbb{R}^n$ then one easily checks that $\mathbb{E}$ is an $\mathcal{A}$–reducing subspace. Thus $\mathbb{E}$ is an $\mathcal{A}$–reducing subspace of $L^2(\mathbb{R}^n)$ if and only if there exists a Lebesgue measurable $\mathcal{A}^*$–set $G \subseteq \mathbb{R}^n$ such that for any $f \in \mathbb{E}$ it follows that $\operatorname{supp} \widehat{f} \subseteq G$ and, moreover, any $g \in L^2(\mathbb{R}^n)$ such that $\operatorname{supp} \widehat{g} \subseteq G$ belongs to the subspace $\mathbb{E}$.

Let $A : \mathbb{R}^n \to \mathbb{R}^n$ be a given linear invertible map which satisfies the conditions (*) and let $G \subset \mathbb{R}^n$ be a given $A^*$-set. Let $H_G^2$ denote the closed linear subspace of $L^2(\mathbb{R}^n)$ defined by

$$H_G^2 = \{f \in L^2(\mathbb{R}^n) : \mathrm{supp}\widehat{f} \subseteq G\}.$$

Note that by Plancherel's theorem $L^2(\mathbb{R}^n)$ is the direct sum of the spaces $H_G^2$ and $H_{G^c}^2$.

It should be mentioned that relevant work connected with the description of the $\mathcal{A}$-reducing subspaces had been done by X. Dai, S. Lu [16], X. Dai, D. Larson, D. M. Speegle [15], [14], E. Hernández, X. Wang, G. Weiss [26], Q. Gu, D. Han [22], X. Dai, Y. Diao, Q. Gu, D. Han [12].

### 2.1. $\mathcal{A}$*−invariant sets*

In this subsection we study $\mathcal{A}$−invariant sets in a somewhat more general context. Let $\mathcal{A} : \mathbb{R}^n \to \mathbb{R}^n$ is a given linear invertible map. Then the following lemma holds.

**Lemma 2.1.** *Let $G \subset \mathbb{R}^n$ be an $\mathcal{A}$−set. Then the following properties hold:*

*(a) $G^c = \mathbb{R}^n \setminus G$ is an $\mathcal{A}$−set;*
*(b) $G$ is an $\mathcal{A}^{-1}$−set;*
*(c) $\overline{G}$ is also an $\mathcal{A}$−set.*

**Proof.** We shall only verify the condition (c). We show that $\overline{G} \subset \mathcal{A}(\overline{G})$. Let $\mathbf{x} \in \overline{G}$, then we take $\{\mathbf{y}_k\} \subset G$ such that $\lim_{k\to\infty} \mathbf{y}_k = \mathbf{x}$. From the continuity of $\mathcal{A}^{-1}$, $\lim_{k\to\infty} \mathcal{A}^{-1}(\mathbf{y}_k) = \mathcal{A}^{-1}(\mathbf{x})$ and by (b) of this lemma, $\mathcal{A}^{-1}(\mathbf{y}_k) \in G$. Thus $\mathcal{A}^{-1}(\mathbf{x}) \in \overline{G}$, and hence $\mathbf{x} \in \mathcal{A}(\overline{G})$. In an analogous way it can be shown that $\mathcal{A}(\overline{G}) \subset \overline{G}$. □

The proof of the following lemma is obvious.

**Lemma 2.2.** *Let $\mathcal{A} : \mathbb{R}^n \to \mathbb{R}^n$ be a linear invertible map. Then for any set $E \subset \mathbb{R}^n$, the set $\Omega_E = \bigcup_{j=-\infty}^{\infty} \mathcal{A}^j E$ is an $\mathcal{A}$−set.*

**Lemma 2.3.** *The set of all $\mathcal{A}$−sets is a $\sigma$−algebra.*

**Proof.** Obviously, $\emptyset$ and $\mathbb{R}^n$ are $\mathcal{A}$−sets. Suppose that we are given a sequence of $\mathcal{A}$−sets, $E_i \subset \mathbb{R}^n$, $i = 1, ...$, we show that $\bigcup_{i=1}^{\infty} E_i$ is an $\mathcal{A}$−set. First we check that $\mathcal{A}(\bigcup_{i=1}^{\infty} E_i) \subseteq \bigcup_{i=1}^{\infty} E_i$. Let $\mathbf{x} \in \mathcal{A}(\bigcup_{i=1}^{\infty} E_i) = \bigcup_{i=1}^{\infty} \mathcal{A}E_i$, then $\mathbf{x} \in \mathcal{A}E_i = E_i$, $i \in \mathbb{N}$, hence $\mathbf{x} \in \bigcup_{i=1}^{\infty} E_i$. The proof of the inclusion $\bigcup_{i=1}^{\infty} E_i \subset \mathcal{A}(\bigcup_{i=1}^{\infty} E_i)$ is verified in a similar way. Finally by Lemma 2.1 if $E$ is an $\mathcal{A}$−set then $E^c$ is an $\mathcal{A}$−set. □

It is easy to see that for any linear invertible map $\mathcal{A}:\mathbb{R}^n\to\mathbb{R}^n$, there exists a non-trivial $\mathcal{A}-$set (different from $\emptyset$ and $\mathbb{R}^n$). By Lemma 2.2 it follows that for any finite set $E\subset\mathbb{R}^n$, the set $\Omega_E=\bigcup_{j=-\infty}^{\infty}\mathcal{A}^jE$ is an $\mathcal{A}-$set. In fact a stronger result holds.

**Theorem 2.1.** *There exists a Lebesgue measurable $\mathcal{A}-$set $G\subset\mathbb{R}^n$, such that $|G|_n>0$ and $|\mathbb{R}^n\setminus G|_n>0$.*

**Proof.** Let $\mathcal{U}=\{\mathbf{u}_i\}_{i=1}^n\subset\mathbb{R}^n$ be a Jordan basis for the map $\mathcal{A}$. Let $\{\mathbf{u}_i\}_{i=1}^k$, $k\le n$, be first $k$ elements of the base $\mathcal{U}$ to which corresponds a Jordan block of size $k\times k$:

$$J_k(\lambda)=\begin{pmatrix}\lambda & 1 & 0 & \dots & 0 & 0\\ 0 & \lambda & 1 & \dots & 0 & 0\\ \dots & \dots & \dots & \dots & \dots & \dots\\ 0 & 0 & 0 & \dots & \lambda & 1\\ 0 & 0 & 0 & \dots & 0 & \lambda\end{pmatrix}, \tag{6}$$

or

$$\widetilde{J}_k(\alpha,\beta)=\begin{pmatrix}\alpha & \beta & 1 & 0 & 0 & \dots & 0 & 0 & 0 & 0\\ -\beta & \alpha & 0 & 1 & 0 & \dots & 0 & 0 & 0 & 0\\ \dots & \dots & \dots & \dots & \dots & \dots & \dots & \dots & \dots & \dots\\ 0 & 0 & 0 & 0 & 0 & \dots & \alpha & \beta & 1 & 0\\ 0 & 0 & 0 & 0 & 0 & \dots & -\beta & \alpha & 0 & 1\\ 0 & 0 & 0 & 0 & 0 & \dots & 0 & 0 & \alpha & \beta\\ 0 & 0 & 0 & 0 & 0 & \dots & 0 & 0 & -\beta & \alpha\end{pmatrix}, \tag{7}$$

where $\lambda$, $\alpha$ and $\beta$ are real numbers.

If the corresponding Jordan block is (6) we put

$$S=\{\mathbf{x}\in\mathbb{R}^n\ :\ \mathbf{x}=\sum_{i=1}^n c_i\mathbf{u}_i\quad\text{and}\quad |c_k|<1\}.$$

If $|\lambda|>1$ it is easy to check that the set $E=\mathcal{A}(S)\setminus S$ has positive Lebesgue measure and for any $l,m\in\mathbb{Z}$, $l\neq m$, $\mathcal{A}^l(E)\cap\mathcal{A}^m(E)=\emptyset$. Hence, if we take any Lebesgue measurable set $F\subset E$ such that $|F|_n>0$ and $|E\setminus F|_n>0$, then the set $\Omega_F=\cup_{j=-\infty}^{\infty}\mathcal{A}^j(F)$ will satisfy to all requirements of Theorem 2.1.

Suppose now that $|\lambda|<1$. Observe that $\lambda\neq 0$, since otherwise the map will not be invertible. Now we let $E=S\setminus\mathcal{A}(S)$ and finish the proof as in the previous case.

If $|\lambda|=1$ then we observe that $\mathcal{A}(S)=S$.

Finally when the corresponding Jordan block is (7) instead of the set $S$ we consider the set

$$\widetilde{S}=\{\mathbf{x}\in\mathbb{R}^n\ :\ \mathbf{x}=\sum_{i=1}^n c_i\mathbf{u}_i\quad\text{and}\quad |c_{k-1}|^2+|c_k|^2<1\}$$

and consider following three cases $\alpha^2+\beta^2>1$, $\alpha^2+\beta^2<1$ and $\alpha^2+\beta^2=1$. The details are left to the reader. □

The polar cone of a set $G$ is defined by $\widehat{G} = \{\mathbf{y} : \mathbf{yx} \leq 0 \quad \text{for any} \quad \mathbf{x} \in G\}$.

**Lemma 2.4.** *Let $G \subset \mathbb{R}^n$ be a non empty $\mathcal{A}-$set, then $\widehat{G}$ is an $\mathcal{A}^*-$set, where $\mathcal{A}^*$ is the adjoint of $\mathcal{A}$.*

**Proof.** That $\mathcal{A}^*(\widehat{G}) \subseteq \widehat{G}$ is evident. Let us show that $\mathcal{A}^*(\widehat{G}) \supseteq \widehat{G}$. For any $\mathbf{y} \in \widehat{G}$ we have that $0 \geq \mathbf{yx}$ for all $\mathbf{x} \in G = \mathcal{A}(G)$, hence $\mathbf{y} \cdot \mathcal{A}^{-1}\mathbf{x} \leq 0$ for all $\mathbf{x} \in G$. Thus $(\mathcal{A}^{-1})^*\mathbf{y} \in \widehat{G}$ concluding the proof. □

**Proposition 2.1.** *Let $\mathcal{A} : \mathbb{R}^n \to \mathbb{R}^n$ be a self-adjoint linear invertible map. If $\mathcal{A}$ has a positive eigenvalue, and $\mathbf{x}_\lambda \in \mathbb{R}^n$ is the corresponding eigenvector, then the polar cone of the singleton $E = \{\mathbf{x}_\lambda\}$ is an $\mathcal{A}-$set.*

**Proof.** By Lemma 2.2 the set $G_E = \cup_{j=-\infty}^{\infty} \mathcal{A}^j E$ is an $\mathcal{A}-$set. Thus, by Lemma 2.4, the polar cone $\widehat{G}_E = \widehat{E}$ is an $\mathcal{A}^*-$set. Hence $\widehat{E}$ is an $\mathcal{A}-$set because $\mathcal{A}$ is a self-adjoint map. □

We will denote by $B_r(\mathbf{y})$ the ball centered at $\mathbf{y}$ with radius $r$. If $\mathbf{y}$ is the origin we simply write $B_r$.

**Definition 2.2.** Let $A : \mathbb{R}^n \to \mathbb{R}^n$ be a linear invertible expansive map and let $G \subset \mathbb{R}^n$ be a Lebesgue measurable set of positive measure. We shall say that a point $\mathbf{z} \in \mathbb{R}^n$ has a $(G, A)-$density number $\mathrm{dn}_A(G, \mathbf{z}) = \rho$ if for any $r > 0$

$$\lim_{j\to\infty} \frac{|G \cap (A^{-j}B_r + \mathbf{z})|_n}{|A^{-j}B_r|_n} = \rho.$$

We will simply write $\mathrm{dn}_A(G)$ if $\mathbf{z} = \mathbf{0}$. If for some $\mathbf{z} \in \mathbb{R}^n$ and $G \subset \mathbb{R}^n$, $|G|_n > 0$, the $(G, A)-$density number $\mathrm{dn}_A(G, \mathbf{z})$ exists then, evidently, $0 \leq \mathrm{dn}_A(G, \mathbf{z}) \leq 1$. When $\mathrm{dn}_A(G, \mathbf{z}) = 1$ we will say that $\mathbf{z}$ is a point of $A-$density for the set $G$ (cf. [11]).

**Lemma 2.5.** *Let $A : \mathbb{R}^n \to \mathbb{R}^n$ be a linear invertible expansive map and let $G \subset \mathbb{R}^n$ be an Lebesgue measurable $A-$set. Then the following properties hold:*

*(a) $\forall G \subset \mathbb{R}^n$ such that $G$ is an $A-$set then $|G|_n = \infty$ or $|G|_n = 0$;*
*(b) the origin has an $(G, A)-$density number $dn_A(G)$ and $dn_A(G^c) = 1 - dn_A(G)$;*
*(c) If $dn_A(G) = 1$ then $|G^c|_n = 0$.*

**Proof.** We shall verify only the condition (b). For any $r > 0$ we have

$$\lim_{j\to\infty} \frac{|G \cap A^{-j}B_r|_n}{|A^{-j}B_r|_n} = \lim_{j\to\infty} \frac{|A^{-j}G \cap A^{-j}B_r|_n}{|A^{-j}B_r|_n}$$

$$= \lim_{j\to\infty} \frac{|A^{-j}(G \cap B_r)|_n}{|A^{-j}B_r|_n} = \frac{|(G \cap B_r)|_n}{|B_r|_n}.$$

Hence,

$$\mathrm{dn}_A(G) = \frac{|(G \cap B_r)|_n}{|B_r|_n} \qquad \text{for any} \quad 0 < r < +\infty. \tag{8}$$

The equality $\mathrm{dn}_A(G^c) = 1 - \mathrm{dn}_A(G)$ plainly follows from (8). The rest of the conditions can be checked in a similar way. □

Using condition (8) we obtain the following interesting fact.

**Lemma 2.6.** *Let $A : \mathbb{R}^n \to \mathbb{R}^n$ be a linear invertible expansive map and suppose that $G_0 \subset G \subset \mathbb{R}^n$ are Lebesgue measurable $A$–sets such that $dn_A(G_0) = dn_A(G)$. Then these sets coincide modulo a null set: $|G \setminus G_0|_n = 0$.*

**Proof.** We have that for any $r > 0$

$$\frac{|(G \setminus G_0) \cap B_r|_n}{|B_r|_n} = \frac{|(G \cap B_r)|_n}{|B_r|_n} - \frac{|(G_0 \cap B_r)|_n}{|B_r|_n} = \mathrm{dn}_A(G) - \mathrm{dn}_A(G_0) = 0. \quad \square$$

## 3. Characterization of Scaling Functions of an FMRA in $H^2_G$

### 3.1. *Definitions and Preliminary results*

Let $\mathbb{T}^n = \mathbb{R}^n/\mathbb{Z}^n$. When we write $F \in L^2(\mathbb{T}^n)$ we will understand that $F$ is defined on the whole space $\mathbb{R}^n$ as 1–periodic function with respect to all variables. With some abuse of the notation we consider also that $\mathbb{T}^n$ is the unit cube $[0,1)^n$. Given a set $E \subset \mathbb{R}^n$ and a real number $a \in \mathbb{R}$, we set $E_{\mathbb{T}^n} = (E + \mathbb{Z}^n)/\mathbb{Z}^n \subseteq \mathbb{T}^n$ and $aE = \{\mathbf{x} \in \mathbb{R}^n : \mathbf{x} = a\mathbf{t} \quad \text{for} \quad \mathbf{t} \in E\}$. We also use the following notations $\mathbf{x} + E = \{\mathbf{x} + \mathbf{y} : \quad \text{for any} \quad \mathbf{y} \in E\}$, where $\mathbf{x} \in \mathbb{R}^n$. The Lebesgue measure of a measurable set $E \subset \mathbb{R}^n$ will be denoted by $|E|_n$. We will write $2^{\mathbf{l}} = (2^{\ell_1}, 2^{\ell_2}, \ldots, 2^{\ell_n}) \in \mathbb{Z}^n$ and $\mathbf{v} = [\mathbf{m}2^{\mathbf{l}}] = (m_1 2^{\ell_1}, m_2 2^{\ell_2}, \ldots, m_n 2^{\ell_n})$, where $\mathbf{m} = (m_1, m_2, \ldots, m_n) \in \mathbb{Z}^n$ and $\mathbf{l} = (\ell_1, \ell_2, \ldots, \ell_n) \in \mathbb{Z}^n$.

**Definition 3.1.** Let $f : \mathbb{R}^n \longrightarrow \mathbb{C}$ be a measurable function. We say that $\mathbf{x} \in \mathbb{R}^n$ is a point of approximate continuity of the function $f$ if there exists $E \subset \mathbb{R}^n$, $|E|_n > 0$, such that $\mathbf{x}$ is a point of density for the set $E$ and

$$\lim_{\substack{\mathbf{y} \to \mathbf{x} \\ \mathbf{y} \in E}} f(\mathbf{y}) = f(\mathbf{x}). \tag{9}$$

It is well known (e.g. [33], [6]) that for any finite measurable function almost all points are points of approximate continuity.

**Definition 3.2.** A measurable function $f : \mathbb{R}^n \to \mathbb{C}$ is said to be *locally nonzero* at a point $\mathbf{x} \in \mathbb{R}^n$ if for any $\varepsilon > 0$, there exists $r$, $0 < r < 1$, such that

$$|\{\mathbf{y} \in B_r(\mathbf{x}) : f(\mathbf{y}) = 0\}|_n < \varepsilon |B_r(\mathbf{x})|_n.$$

In Sections 3 and 4 we will suppose that $A$ is a given linear invertible map $A : \mathbb{R}^n \to \mathbb{R}^n$ which satisfies the conditions (*). The reader should note that several definitions that will be given below make sense without the condition $A\mathbb{Z}^n \subset \mathbb{Z}^n$.

**Definition 3.3.** We will say that $\mathbf{x} \in \mathbb{R}^n$ is a point of $A$–density for a set $E \subset \mathbb{R}^n$, $|E|_n > 0$ if for any $r > 0$,

$$\lim_{j\to\infty} \frac{|E \cap (A^{-j}B_r + \mathbf{x})|_n}{|A^{-j}B_r|_n} = 1.$$

**Definition 3.4.** Let $f : \mathbb{R}^n \longrightarrow \mathbb{C}$ be a measurable function. We say that $\mathbf{x} \in \mathbb{R}^n$ is a point of $A$–approximate continuity of the function $f$ if there exists $E \subset \mathbb{R}^n$, $|E|_n > 0$, such that $\mathbf{x}$ is a point of $A$–density for the set $E$ and (9) holds.

**Definition 3.5.** A measurable function $f : \mathbb{R}^n \to \mathbb{C}$ is said to be $A$–*locally nonzero* at a point $\mathbf{x} \in \mathbb{R}^n$ if for any $\varepsilon > 0$ and $r > 0$ there exists $j \in \mathbb{N}$ such that

$$|\{\mathbf{y} \in A^{-j}B_r + \mathbf{x} : f(\mathbf{y}) = 0\}|_n < \varepsilon |A^{-j}B_r|_n.$$

Now we are ready to formulate the result obtained in [11].

**Theorem J.** *Let $V_j$ be a sequence of closed subspaces in $L^2(\mathbb{R}^n)$ such that the conditions (i), ($ii_1$) and (iv) hold. Then the following conditions are equivalent:*

(*a*) $W = \overline{\cup_{j\in\mathbb{Z}} V_j} = L^2(\mathbb{R}^n)$;
(*b*) $\widehat{\phi}$ *—the Fourier transform of the scaling function* $\phi$*— is* $A^*$*–locally nonzero at the origin*;
(*c*) *the origin is a point of* $A^*$*–approximate continuity of the function* $|\widehat{\phi}|$ *if we set* $|\widehat{\phi}(\mathbf{0})| = 1$.

**Remark 3.1.** If in Theorem J one changes the condition (*b*) by the condition:

$$(b^*) \qquad \liminf_{\substack{j\to+\infty \\ r\to+\infty}} \frac{|\{\mathbf{y}\in A^{-j}B_r+\mathbf{x}: f(\mathbf{y})=0\}|_n}{|A^{-j}B_r|_n} = 0,$$

then the theorem will remain true.

The above remark follows from Theorem 3.1 which will be proved in Section 3.

Since the beginning of the theory of wavelets the conditions used in the definition of an MRA have been modified in order to attend different purposes required by applications. In particular this is the case with the condition (iv). The requirement that $\{\phi(\mathbf{x}-\mathbf{k})\}_{\mathbf{k}\in\mathbb{Z}^n}$ is an orthonormal basis for $V_0$ was weakened at first, requiring that the shifts of the scaling functions constitute a Riesz basis for $V_0$. Afterwards, the notion of a *frame multiresolution analysis* (FMRA) for $L^2(\mathbb{R})$ was introduced by J. Benedetto and S. Li [1]. An $A$-FMRA is a natural extension of the $A$-MRA that is obtained by replacing the condition (iv) with the following one:

(iv)* There exists a function $\phi \in V_0$, that is called *scaling function*, such that $\{\phi(\mathbf{x}-\mathbf{k})\}_{\mathbf{k}\in\mathbb{Z}^n}$ is a frame for $V_0$.

The theory of frames was introduced by R. Duffin and A. Schaeffer [20].

**Definition 3.6.** A sequence $\{\phi_n\}_{n=1}^{\infty}$ of elements in a separable Hilbert space $\mathbb{H}$ is a *frame* for $\mathbb{H}$ if there exist constants $B, C > 0$ such that

$$B\|h\|^2 \leq \sum_{n=1}^{\infty} |\langle h, \phi_n \rangle|^2 \leq C\|h\|^2, \qquad \forall h \in \mathbb{H}.$$

The constants $B$ and $C$ are called *frame bounds.* It is clear that a frame is a complete set of elements in $\mathbb{H}$, since the relations $\langle h, \phi_n \rangle = 0, n \in \mathbb{N}$, imply that $h = 0$. A frame $\{\phi_n\}_{n=1}^{\infty}$ is *tight* if we can choose $C = B$; and is called *normalized tight frame* if $C = B = 1$. A frame $\{\phi_n\}_{n=1}^{\infty}$ is *exact* if it ceases to be a frame when any one of its elements is removed.

If $\{\phi_n\}_{n=1}^{\infty}$ is a frame for $\mathbb{H}$ then the *frame operator* $S : \mathbb{H} \to \mathbb{H}$ is the bounded linear operator defined by

$$Sh = \sum_{n=1}^{\infty} \langle h, \phi_n \rangle \phi_n.$$

It can be shown that the operator $S$ is self-adjoint, positive and invertible (e.g. [20] , [17], [9]) and that $\{S^{-1}\phi_n\}_{n=1}^{\infty}$ is a frame for $\mathbb{H}$. Moreover, $\forall h \in \mathbb{H}$

$$h = SS^{-1}h = \sum_{n=1}^{\infty} \langle S^{-1}h, \phi_n \rangle \phi_n = \sum_{n=1}^{\infty} \langle h, S^{-1}\phi_n \rangle \phi_n.$$

The following theorem explains the close relation that exists between a normalized tight frame and an orthonormal system (see [23]).

**Theorem K.** *Let $\{\phi_n\}_{n=1}^{\infty}$ be a normalized tight frame for a Hilbert space $\mathbb{H}$. Then there exists a Hilbert space $\mathbb{F} \supseteq \mathbb{H}$ and an orthonormal basis $\{e_n\}_{n=1}^{\infty}$ for $\mathbb{F}$ such that $\phi_n = Pe_n$ for any $n \in \mathbb{N}$, where $P$ is the orthogonal projection from $\mathbb{F}$ onto $\mathbb{H}$.*

From Theorem K follows that when $\{\phi_n\}_{n=1}^{\infty}$ is a normalized tight frame for $\mathbb{H}$ then $\forall h \in \mathbb{H}$

$$h = \sum_{n=1}^{\infty} \langle h, \phi_n \rangle \phi_n. \tag{10}$$

**Definition 3.7.** A sequence $\{h_n\}_{n=1}^{\infty}$ of elements in a Hilbert space $\mathbb{H}$ is a *frame sequence* if it is a frame for $\overline{\text{span}}\{h_n\}_{n=1}^{\infty}$.

Following C. de Boor, R. DeVore and A. Ron [4] we will use the concept of the bracket product which is defined as follows: for $f, g \in L^2(\mathbb{R}^n)$ we put

$$[f, g](\cdot) = \sum_{\mathbf{k} \in \mathbb{Z}^n} f(\cdot + \mathbf{k})\bar{g}(\cdot + \mathbf{k}). \tag{11}$$

Let

$$\mathcal{N}_\phi = \{\mathbf{t} \in \mathbb{T}^n : [\widehat{\phi}, \widehat{\phi}](\mathbf{t}) = 0\} \tag{12}$$

and will consider that $\frac{0}{0} = 0$ or $0\frac{1}{0} = 0$ in some expressions where such an indeterminacy appears.

The following theorem was proved in 1992 by J. J. Benedetto and S. Li (cf. [1]) on $\mathbb{R}$ and an extension to $\mathbb{R}^n$ was given in [2]. Independently C. de Boor, R. A. De Vore and A. Ron [3] proved the result for finitely generated shift invariant spaces in terms of their notion of quasi-stable bases (cf. [36]).

**Theorem L.** *Let $\phi \in L^2(\mathbb{R}^n)$ and let $V = \overline{span}\{T_{\mathbf{k}}\phi \,:\, \mathbf{k} \in \mathbb{Z}^n\}$ be a closed subspace of $L^2(\mathbb{R}^n)$. Assume $[\phi,\phi] \in L^\infty(\mathbb{T}^n)$. The sequence $\{T_{\mathbf{k}}\phi \,:\, \mathbf{k} \in \mathbb{Z}^n\}$ is a frame for $V$ if and only if there are positive constants $B$ and $C$ such that*

$$B \leq [\phi,\phi](\mathbf{t}) \leq C \qquad a.e.\ on\ \mathbb{T}^n \setminus \mathcal{N}_\phi.$$

*In this case $B$ and $C$ are frame bounds for $\{T_{\mathbf{k}}\phi \,:\, \mathbf{k} \in \mathbb{Z}^n\}$.*

In an implicit form the following proposition can be found in [3, Theorem 2.21] (cf. [1, 2]. It gives the characterization of scaling functions $\phi$, when in the definition of an MRA the condition (iv) is replaced by the following one:

(iv)** $$V_0 = \overline{\text{span}}\{\phi(\mathbf{x}-\mathbf{k}) \,:\, \mathbf{k} \in \mathbb{Z}^n\}.$$

**Proposition A.** Let $\phi \in L^2(\mathbb{R}^n)$ and let $V = \overline{\text{span}}\{T_{\mathbf{k}}\phi \,:\, \mathbf{k} \in \mathbb{Z}^n\}$. Then the system $\{T_{\mathbf{k}}\varphi \,:\, \mathbf{k} \in \mathbb{Z}^n\}$, where the function $\varphi$ is defined via its Fourier transform by (21) is a normalized tight frame for $V$.

In [28] and [10] characterizations of scaling functions for dyadic frame multiresolution analyses were given. H. O. Kim, R. Y. Kim and J. K. Lim [28] generalized Theorem H for an FMRA in $L^2(\mathbb{R})$.

### 3.2. *Characterization of scaling functions of an $H^2_G$-FMRA and other cases*

For a given linear invertible map $A : \mathbb{R}^n \to \mathbb{R}^n$ which satisfies to the conditions (*) and for a measurable $A^*$−set $G \subseteq \mathbb{R}^n, |G|_n > 0$, we will consider an $H^2_G$-MRA associated with $A$ as a sequence of closed subspaces $V_j$, $j \in \mathbb{Z}$ of the Hilbert space $H^2_G$ that satisfies the following conditions:

($\mathrm{i}_G$) $\forall j \in \mathbb{Z}, \quad V_j \subset V_{j+1} \subset H^2_G$;

($\mathrm{ii}_G$) $\forall j \in \mathbb{Z}, \quad f(\mathbf{x}) \in V_j \Leftrightarrow f(A\mathbf{x}) \in V_{j+1}$ ;

($\mathrm{iii}_G$) $W_G = \overline{\cup_{j\in\mathbb{Z}} V_j} = H^2_G$;

($\mathrm{iv}_G$) There exists a function $\phi \in V_0$, that is called *scaling function*, such that the system $\{\phi(\mathbf{x}-\mathbf{k})\}_{\mathbf{k}\in\mathbb{Z}^n}$ is an orthonormal basis for $V_0$.

If the condition ($\mathrm{iv}_G$) is replaced by

($\mathrm{iv}_G$)* the system $\{\phi(\mathbf{x}-\mathbf{k})\}_{\mathbf{k}\in\mathbb{Z}^n}$ is a frame for $V_0$,

then we will say that we are considering an $H^2_G$-FMRA associated with $A$.

**Definition 3.8.** We will say that a function $\phi \in H^2_G$ *generates an $H^2_G$-FMRA associated with $A$* if $\{T_{\mathbf{k}}\phi\}_{\mathbf{k}\in\mathbb{Z}^n}$ is a frame sequence and the subspaces

$$V_j = \overline{\operatorname{span}}\{D^j_A T_{\mathbf{k}}\phi\}_{\mathbf{k}\in\mathbb{Z}^n}, \quad j \in \mathbb{Z} \tag{13}$$

of the Hilbert subspace $H^2_G$ satisfy the conditions ($\mathrm{i}_G$) and ($\mathrm{iii}_G$).

We are going to characterize those functions $\phi$ which generate an $H^2_G$-FMRA associated with $A$. For that purpose we need some new definitions.

**Definition 3.9.** Let $A : \mathbb{R}^n \to \mathbb{R}^n$ be a linear invertible expansive map and let $G \subset \mathbb{R}^n$ be a measurable $A$–set, $|G| > 0$, and let $E \subset \mathbb{R}^n$ be any measurable set with non vanishing Lebesgue measure. We will say that $\mathbf{x} \in \mathbb{R}^n$, is a $(G, A)$–density point of the set $E$ if $\mathrm{dn}_A(E \bigcap (G + \mathbf{x}), \mathbf{x}) = \mathrm{dn}_A(G)$.

**Definition 3.10.** Let $G \subset \mathbb{R}^n$ be an $A$–set, $|G| > 0$, and let $f : \mathbb{R}^n \longrightarrow \mathbb{C}$ be a measurable function. We say that $\mathbf{x} \in \mathbb{R}^n$ is a point of $(G, A)$-approximate continuity of the function $f$ if there exists a measurable set $E$ such that $\mathbf{x}$ is a $(G, A)$–density point of the set $E$ and the condition (9) holds.

**Definition 3.11.** Let $G \subset \mathbb{R}^n$ be an $A$–set, $|G|_n > 0$. A measurable function $f : \mathbb{R}^n \to \mathbb{C}$ is said to be $(G, A)$-*locally nonzero at a point* $\mathbf{x} \in \mathbb{R}^n$ if

$$\liminf_{\substack{j\to+\infty \\ r\to+\infty}} \frac{|\{\mathbf{y} \in (A^{-j}B_r + \mathbf{x}) \cap (G + \mathbf{x}) : f(\mathbf{y}) = 0\}|_n}{|(A^{-j}B_r) \cap G|_n} = 0.$$

We prove the following

**Theorem 3.1.** *Let $\phi \in H^2_G$. Then the following conditions are equivalent:*

*(**A**) $\phi$ generates an $H^2_G$-FMRA associated with $A$;*
*(**B**) ($\alpha$) The function $\widehat{\phi}$ is $(G, A^*)$-nonzero at the origin;*
*($\beta$) there exist positive constants $B$, $C$ such that*

$$B \leq [\widehat{\phi}, \widehat{\phi}](\mathbf{t}) \leq C \quad \text{a.e. on} \quad \mathbb{T}^n \setminus \mathcal{N}_\phi; \tag{14}$$

*($\gamma$) there exists $H \in L^\infty(\mathbb{T}^n)$, which is called low pass filter, such that*

$$\widehat{\phi}(\mathbf{t}) = H((A^*)^{-1}\mathbf{t})\widehat{\phi}((A^*)^{-1}\mathbf{t}) \qquad \text{a.e. on } \mathbb{R}^n; \tag{15}$$

*(**C**) ($\alpha'$) The origin is a point of $(G, A^*)$-approximate continuity of the function $|\widehat{\phi}|^2 \cdot ([\widehat{\phi}, \widehat{\phi}])^{-1}$, provided that $|\widehat{\phi}(\mathbf{0})|^2([\widehat{\phi}, \widehat{\phi}](\mathbf{0}))^{-1} = 1$, and conditions ($\beta$) and ($\gamma$) hold.*

From Theorem 3.1 it immediately follows that

**Claim 3.1.** *If $0 < dn_{A^*}(G) < 1$, and $\phi \in H^2_G$ generates an $H^2_G$-FMRA associated with $A$, then the function $\phi \notin L^1(\mathbb{R}^n)$.*

For the proof of Theorem 3.1 we establish a series of lemmas. Different versions of Lemmas 3.1–3.2 formulated below have appeared in various publications (cf. [17], pp. 131–132; [38], p. 28–29, [25] pp. 381–382 and [39] ).

**Lemma 3.1.** *Let $\phi \in L^2(\mathbb{R}^n)$ and assume that $\{T_{\mathbf{k}}\phi\}_{\mathbf{k}\in\mathbb{Z}^n}$ is a frame sequence. Then a function $f$ is in $V_j$, where $V_j$ is defined by (13), if and only if there is an equivalence class $\mathcal{F}_j$ of the factor space $L^2(\mathbb{T}^n)/L^2(\mathcal{N}_\phi)$ such that for any $F \in \mathcal{F}_j$*

$$D^j_{A^*}\widehat{f}(\mathbf{t}) = F(\mathbf{t})\widehat{\phi}(\mathbf{t}) \quad a.e. \text{ on } \quad \mathbb{R}^n \quad \text{if} \quad j \in \mathbb{Z}^n. \tag{16}$$

We leave the proof to the reader, remarking that the key relation for the proof is the equality (2) (see also (1)).

**Lemma 3.2.** *Let $\phi \in L^2(\mathbb{R}^n)$ and assume that $\{T_{\mathbf{k}}\phi\}_{\mathbf{k}\in\mathbb{Z}^n}$ is a frame sequence in $L^2(\mathbb{R}^n)$. If the subspaces $V_j, j \in \mathbb{Z}$, are defined by (13) then the following conditions are equivalent:*

*a*)* $\forall j \in \mathbb{Z}, \quad V_j \subset V_{j+1}$;

*b*)* $V_0 \subset V_1$;

*c*)* *There exists an equivalence class $\mathcal{F}$ of $L^\infty(\mathbb{T}^n)/L^\infty(\mathcal{N}_\phi)$ such that for any $H \in \mathcal{F}$*

$$D_{A^*}\widehat{\phi}(\mathbf{t}) = H(\mathbf{t})\widehat{\phi}(\mathbf{t}) \quad a.e. \text{ on } \quad \mathbb{R}^n.$$

**Proof.** The implication a*)⇒ b*) is obvious. If b*) holds then $D_A^{-1}\phi \in V_{-1}$. By Lemma 3.1 there is $H \in L^2(\mathbb{T}^n)$ such that $\widehat{D_A^{-1}\phi}(\mathbf{t}) = H(\mathbf{t})\widehat{\phi}(\mathbf{t})$. On the other hand we verify directly that $\widehat{D_A^{-1}\phi}(\mathbf{t}) = d_A^{\frac{1}{2}}\widehat{\phi}(A^*\mathbf{t})$. Hence,

$$\widehat{\phi}(A^*\mathbf{t}) = H_0(\mathbf{t})\widehat{\phi}(\mathbf{t}), \qquad \text{where} \quad H_0 \in L^2(\mathbb{T}^n). \tag{17}$$

Redefine $H_0$ to be zero on the set $\mathcal{N}_\phi$. Then

$$[\widehat{\phi},\widehat{\phi}](\mathbf{t}) = \sum_{\mathbf{k}\in\mathbb{Z}^n} |\widehat{\phi}(\mathbf{t}+\mathbf{k})|^2 = \sum_{\mathbf{k}\in A^*(\mathbb{Z}^n)} |\widehat{\phi}(\mathbf{t}+\mathbf{k})|^2$$

$$+ \sum_{\mathbf{k}\notin A^*(\mathbb{Z}^n)} |\widehat{\phi}(\mathbf{t}+\mathbf{k})|^2 = |H_0((A^*)^{-1}\mathbf{t})|^2[\widehat{\phi},\widehat{\phi}]((A^*)^{-1}\mathbf{t}) + R(\mathbf{t})$$

where $R(\mathbf{t}) \geq 0$. Hence, by Theorem L we will have that for any $\mathbf{t} \notin \mathcal{N}_\phi$

$$C \geq [\widehat{\phi},\widehat{\phi}](A^*\mathbf{t}) \geq |H_0(\mathbf{t})|^2[\widehat{\phi},\widehat{\phi}](\mathbf{t}) \geq B|H_0(\mathbf{t})|^2$$

and therefore,

$$|H_0(\mathbf{t})| \leq (C/B)^{\frac{1}{2}}. \tag{18}$$

To prove c*) ⇒ a*) we take any $f \in V_j$. Then, by Lemma 3.1, there is a function $F_f \in L^2(\mathbb{T}^n)$ such that

$$D^j_{A^*}\widehat{f}(\mathbf{t}) = F_f(\mathbf{t})\widehat{\phi}(\mathbf{t}) = F_f(\mathbf{t})H((A^*)^{-1}\mathbf{t})\widehat{\phi}((A^*)^{-1}\mathbf{t}),$$

where the last equation follows from (17).

Hence,

$$D^{j+1}_{A^*}\widehat{f}(\mathbf{t}) = d^{\frac{1}{2}}_{A^*}F_f(A^*\mathbf{t})H(\mathbf{t})\widehat{\phi}(\mathbf{t}),$$

where $F_f(A^*\mathbf{t})H(\mathbf{t}) \in L^2(\mathbb{T}^n)$ and therefore by Lemma 3.1 we get $f \in V_{j+1}$. □

**Remark 3.2.** If in Lemma 3.2 we suppose that $\{T_\mathbf{k}\phi\}_{\mathbf{k}\in\mathbb{Z}^n}$ is an orthonormal basis for $V_0$ then naturally $|\mathcal{N}_\phi| = 0$ and by (18) we would also have that $\|H\|_\infty \leq 1$.

**Lemma 3.3.** *Let $V_j$ be a sequence of closed subspaces in $H^2_G$ satisfying the conditions $(i_G)$, $(ii_G)$, $(iii_G)$ and $(iv_G)^*$. Then for any bounded measurable set $E \subset G$, $|E|_n > 0$,*

$$\lim_{j\to\infty}\frac{1}{|(A^*)^{-j}E|_n}\int_{(A^*)^{-j}E}|\widehat{\phi}(\mathbf{t})|^2([\widehat{\phi},\widehat{\phi}](\mathbf{t}))^{-1}d\mathbf{t} = 1. \tag{19}$$

**Proof.** For any bounded measurable set $E \subset G$, $|E|_n > 0$, we take $f \in H^2_G$ such that $\widehat{f} = \chi_E$, $|E|_n > 0$. Then

$$\|f\|_2^2 = \|\widehat{f}\|_2^2 = |E|_n.$$

Let $P_j$ be the orthogonal projection onto $V_j$. Then, by properties $(\mathrm{i}_G)$ and $(\mathrm{iii}_G)$ we have $\|f - P_jf\|_2 \to 0$ as $j \to \infty$. Hence, when $j \to \infty$,

$$\|P_jf\|_2^2 \to \|f\|_2^2 = |E|_n. \tag{20}$$

If we define $\varphi$ by

$$\widehat{\varphi} = \widehat{\phi}/[\widehat{\phi},\widehat{\phi}]^{1/2}, \tag{21}$$

then by Proposition A it is easy to verify (cf. [10]) that the system $\{\varphi_{j\mathbf{k}}\}_{\mathbf{k}\in\mathbb{Z}^n}$, where

$$\varphi_{j\mathbf{k}}(\mathbf{x}) = d_A^{\frac{j}{2}}\varphi(A^j\mathbf{x} - \mathbf{k}),$$

will be a normalized tight frame for $V_j$. Observe that

$$\widehat{\varphi_{j\mathbf{k}}}(\mathbf{t}) = d_A^{-\frac{j}{2}}e^{-2\pi i\mathbf{k}\cdot(A^*)^{-j}\mathbf{t}}\widehat{\varphi}((A^*)^{-j}\mathbf{t}).$$

Thus, by (10) we will have that

$$P_jf = \sum_{k\in\mathbb{Z}^n}\langle P_jf,\varphi_{j\mathbf{k}}\rangle\varphi_{j\mathbf{k}} = \sum_{k\in\mathbb{Z}^n}\langle f,\varphi_{j\mathbf{k}}\rangle\varphi_{j\mathbf{k}}$$

and

$$\begin{aligned}\|P_jf\|_2^2 &= \sum_{\mathbf{k}\in\mathbb{Z}^n}|\langle f,\varphi_{j\mathbf{k}}\rangle|^2 = \sum_{\mathbf{k}\in\mathbb{Z}^n}\left|\int_{\mathbb{R}^n}f(\mathbf{x})\overline{\varphi_{j\mathbf{k}}(\mathbf{x})}d\mathbf{x}\right|^2\\ &= \sum_{\mathbf{k}\in\mathbb{Z}^n}\left|\int_{\mathbb{R}^n}\widehat{f}(\mathbf{t})\overline{\widehat{\varphi_{j\mathbf{k}}}(\mathbf{t})}d\mathbf{t}\right|^2 = \sum_{\mathbf{k}\in\mathbb{Z}^n}\left|d_A^{-\frac{j}{2}}\int_{\mathbb{R}^n}\widehat{f}(\mathbf{t})e^{2\pi i\mathbf{k}\cdot(A^*)^{-j}\mathbf{t}}\overline{\widehat{\varphi}((A^*)^{-j}\mathbf{t})}d\mathbf{t}\right|^2.\end{aligned}$$

Hence, after the change of variables, we get

$$\|P_j f\|_2^2 = \sum_{\mathbf{k}\in\mathbb{Z}^n} \left| d_A^{\frac{j}{2}} \int_{\mathbb{R}^n} \widehat{f}(A^{*j}\mathbf{y})\overline{\widehat{\varphi}(\mathbf{y})} e^{2\pi i\mathbf{k}\cdot\mathbf{y}} d\mathbf{y} \right|^2$$

$$= \sum_{\mathbf{k}\in\mathbb{Z}^n} \left| d_A^{\frac{j}{2}} \int_{(A^*)^{-j}E} \overline{\widehat{\varphi}(\mathbf{y})} e^{2\pi i\mathbf{k}\cdot\mathbf{y}} d\mathbf{y} \right|^2,$$

where $(A^*)^{-j}E = \{\mathbf{y} \in \mathbb{R}^n : A^{*j}\mathbf{y} \in E\}$. Let $j_1$ be the minimal natural number such that $A^{*-j_1}E \subset [-\frac{1}{2}, \frac{1}{2}]^n$. Then, for any $j \geq j_1$, the last sum is equal to

$$d_A^j \int_{\mathbb{R}^n} |\chi_{(A^*)^{-j}E}(\mathbf{t})\widehat{\varphi}(\mathbf{t})|^2 d\mathbf{t},$$

by Parseval's equality. Therefore, by the definition of the function $\varphi$ and the condition (20), we obtain (19). □

**Proof of Theorem 3.1.** Let us begin with the proof of the implication **(B)** ⇒ **(A)**. Applying Lemma 3.2 and Theorem L we obtain that $\{T_{\mathbf{k}}\phi\}_{\mathbf{k}\in\mathbb{Z}^n}$ is a frame for $V_0$ and $V_j \subset V_{j+1}$ for every $j \in \mathbb{Z}$, where $V_j$ is defined by (13). We have to prove that (iii$_G$) holds. Obviously, $W_G \subseteq H_G^2$. Thus we have to prove that $W_G \supseteq H_G^2$. For that purpose we observe that the closed subspace $W_G$ is an $A$-reducing subspace which according to Remark 2.1 is invariant under translations. As in the proof of Lemma A we will have that for any $g \in W_G^{\perp}$ and any $f \in W_G$ the condition (3) holds.

For any fixed $j \in \mathbb{N}$ if we take $f = D_A^j\phi \in V_j$, then by the equation (2) we will have $\widehat{f}(\mathbf{y}) = D_{A^*}^{-j}\widehat{\phi}$ and $\widehat{\phi}((A^*)^{-j}\mathbf{y})\overline{\widehat{g}(\mathbf{y})} = 0$ a.e. on $\mathbb{R}^n$. Hence,

$$\widehat{\phi}(\mathbf{t})\overline{\widehat{g}((A^*)^j\mathbf{t})} = 0 \quad \text{a.e.} \quad \mathbb{R}^n. \tag{22}$$

Let $r_0 > 1$ be any fixed number. According to our hypothesis, for any positive integer $N > r_0$ there exist $k \in \mathbb{N}$ and $r_N > r_0$ such that

$$|\{\mathbf{t} \in (A^*)^{-k}B_{r_N} \cap G \ : \ \widehat{\phi}(\mathbf{t}) = 0\}|_n < \frac{|(A^*)^{-k}B_{r_N} \cap G|_n}{N},$$

and since $G$ is an $A^*$-set, we have
$(A^*)^{-k}B_{r_N} \cap G = (A^*)^{-k}B_{r_N} \cap (A^*)^{-k}G = (A^*)^{-k}(B_{r_N} \cap G)$.
Hence,

$$|\{\mathbf{t} \in (A^*)^{-k}(B_{r_N} \cap G) \ : \ \widehat{\phi}(\mathbf{t}) = 0\}|_n < \frac{|(A^*)^{-k}(B_{r_N} \cap G)|_n}{N}$$

and by (3)

$$|\{\mathbf{t} \in (A^*)^{-k}(B_{r_N} \cap G) \ : \ \widehat{g}((A^*)^k\mathbf{t}) \neq 0\}|_n < \frac{|(A^*)^{-k}(B_{r_N} \cap G)|_n}{N}$$

and therefore taking $j = k$ we obtain by (22)

$$|\{\mathbf{y} \in B_{r_N} \cap G \ : \ \widehat{g}(\mathbf{y}) \neq 0\}|_n < \frac{|B_{r_N} \cap G|_n}{N}.$$

The numbers $r_N$ have been chosen so that $r_N > r_0$ for any $N > r_0$. Thus we have

$$|\{\mathbf{y} \in B_{r_0} \cap G \ : \ \widehat{g}(\mathbf{y}) \neq 0\}|_n < \frac{|B_{r_N} \cap G|_n}{N}.$$

Letting $N \to \infty$ we obtain $|\{\mathbf{y} \in B_{r_0} \cap G \ : \ \widehat{g}(\mathbf{y}) \neq 0\}|_n = 0$. Hence, $\widehat{g} = 0$ a.e. on $G$, and therefore $g \in H^2_{G^c}$. Therefore we have proved that $W_G^{\perp} \subseteq H^2_{G^c}$ and consequently $W_G \supseteq H^2_G$.

Let us prove the implication (**A**) $\Rightarrow$ (**C**). By Lemma 3.2 and Theorem L we immediately see that conditions (14) and (15) hold. It is obvious that

$$|\widehat{\phi}(\mathbf{t})|^2([\widehat{\phi},\widehat{\phi}](\mathbf{t}))^{-1} \leq 1, \qquad \text{a.e. on } \mathbb{R}^n. \tag{23}$$

We have to show that there exists $E \subset G$, $|E|_n > 0$, such that the origin is a $(G, A^*)$–density point of the set $E$ and

$$\lim_{\substack{\mathbf{y} \to \mathbf{0} \\ \mathbf{y} \in E}} |\widehat{\phi}(\mathbf{y})|^2([\widehat{\phi},\widehat{\phi}](\mathbf{y}))^{-1} = 1.$$

In its turn this is equivalent to the following statement:
for any $\varepsilon > 0$ and any $r > 0$,

$$\lim_{j\to\infty} \frac{|\{\mathbf{y} \in (A^*)^{-j}(B_r \cap G) \ : \ \big||\widehat{\phi}(\mathbf{y})|^2([\widehat{\phi},\widehat{\phi}](\mathbf{y}))^{-1} - 1\big| < \varepsilon\}|_n}{|(A^*)^{-j}(B_r \cap G)|_n} = 1.$$

Suppose that our claim is false. Then, having in mind (23), we obtain that there exist $0 < \varepsilon_0 < 1$, $r_0 > 0$ and an increasing sequence of natural numbers $\{m_j\}_{j=1}^{\infty}$ such that

$$\begin{aligned}|\Gamma_j|_n &= |\{\mathbf{y} \in (A^*)^{-m_j}(B_{r_0} \cap G) \ : \ |\widehat{\phi}(\mathbf{y})|^2([\widehat{\phi},\widehat{\phi}](\mathbf{y}))^{-1} < 1 - \varepsilon_0\}|_n \\ &\geq \varepsilon_0 |(A^*)^{-m_j}(B_{r_0} \cap G)|_n.\end{aligned}$$

By Lemma 3.3 and by condition (23) we have

$$\begin{aligned}
1 &= \lim_{j\to\infty} |(A^*)^{-m_j}(B_{r_0} \cap G)|_n^{-1} \int_{(A^*)^{-m_j}(B_{r_0}\cap G)} |\widehat{\phi}(\mathbf{t})|^2([\widehat{\phi},\widehat{\phi}](\mathbf{t}))^{-1} d\mathbf{t} \\
&= \lim_{j\to\infty} |(A^*)^{-m_j}(B_{r_0} \cap G)|_n^{-1} \\
&\times \left( \int_{[(A^*)^{-m_j}(B_{r_0}\cap G)]\setminus\Gamma_j} |\widehat{\phi}(\mathbf{t})|^2([\widehat{\phi},\widehat{\phi}](\mathbf{t}))^{-1} d\mathbf{t} + \int_{\Gamma_j} |\widehat{\phi}(\mathbf{t})|^2([\widehat{\phi},\widehat{\phi}](\mathbf{t}))^{-1} d\mathbf{t} \right) \\
&\leq \lim_{j\to\infty} \frac{|(A^*)^{-m_j}(B_{r_0} \cap G)|_n - |\Gamma_j|_n + (1-\varepsilon_0)|\Gamma_j|_n}{|(A^*)^{-m_j}(B_{r_0} \cap G)|_n} \\
&\leq \lim_{j\to\infty} \frac{|(A^*)^{-m_j}(B_{r_0} \cap G)|_n - \varepsilon_0^2 |(A^*)^{-m_j}(B_{r_0} \cap G)|_n}{|(A^*)^{-m_j}(B_{r_0} \cap G)|_n} \leq 1 - \varepsilon_0^2.
\end{aligned}$$

This contradiction concludes the proof of this implication.

Since the implication (**C**) $\Rightarrow$ (**B**) is trivial the proof of Theorem 3.1 is now complete. □

The unique piece which one needs to add to Theorem 3.1 to obtain the characterization of scaling functions of an $H^2_G$-MRA associated with $A$ is the following well known lemma (cf. [25, p. 50], [39, p. 111] ).

**Lemma 3.4.** *The system $\{g(\cdot - \mathbf{k}) : \mathbf{k} \in \mathbb{Z}^n\}$, where $g \in L^2(\mathbb{R}^n)$, is an orthonormal system if and only if*

$$\sum_{\mathbf{k}\in\mathbb{Z}^n} |\widehat{g}(\mathbf{t}+\mathbf{k})|^2 = 1 \qquad \textit{for a.e.} \quad \mathbf{t} \in \mathbb{R}^n.$$

The following result is true.

**Corollary 3.1.** *Let $\phi \in H^2_G$. Then the following conditions are equivalent:*

*($\mathbf{A}_2$) $\phi$ is an scaling function for an $H^2_G$-MRA associated with A;*
*($\mathbf{B}_2$) (a) The function $\widehat{\phi}$ is $(G, A^*)$-locally nonzero at the origin;*
*(b) $[\widehat{\phi}, \widehat{\phi}](\mathbf{t}) = 1$ a. e. on $\mathbb{T}^n$;*
*(c) there exists $H \in L^\infty(\mathbb{T}^n)$, $\|H\|_\infty \leq 1$, a $\mathbb{Z}^n$–periodic function, which is called low pass filter, for which*

$$\widehat{\phi}(\mathbf{t}) = H((A^*)^{-1}\mathbf{t})\widehat{\phi}((A^*)^{-1}\mathbf{t}) \qquad \textit{a.e. on } \mathbb{R}^n; \tag{24}$$

*($\mathbf{C}_2$) The origin is a point of $(G, A^*)$-approximate continuity of the function $|\widehat{\phi}|$, provided that $|\widehat{\phi}(\mathbf{0})| = 1$, and that the conditions (b) and (c) hold.*

One should observe that the estimate $\|H\|_\infty \leq 1$ in the above result follows by Remark 3.2.

Multiresolution analysis generated by a scaling function $\phi$, where the condition (iv) is replaced by the requirement that $V_0$ is the closed linear span in $L^2(\mathbb{R}^n)$ of $\{\phi(\mathbf{x}-\mathbf{k})\}_{\mathbf{k}\in\mathbb{Z}^n}$, has been studied by several authors (cf. [4], [3], [37], [18], [10]). Observe that Theorem E corresponds to the above mentioned case.

If in the definition of an $H^2_G$ we replace the condition (iv$_G$) by the following one:

$$(\text{iv}_G)^{**} \qquad V_0 = \overline{\text{span}}\{\phi(\mathbf{x}-\mathbf{k}) \ : \ \mathbf{k} \in \mathbb{Z}^n\},$$

then by Theorem 3.1 and Proposition A we obtain the following result.

**Corollary 3.2.** *Let $\phi \in H^2_G$. Then the following conditions are equivalent:*

*($\mathbf{A}_3$) The conditions (i$_G$), (ii$_G$), (iii$_G$) and (iv$_G$)$^{**}$ hold;*
*($\mathbf{B}_3$) The function $\widehat{\phi}$ is $(G, A^*)$-locally nonzero at the origin and there exists a $\mathbb{Z}^n$-periodic function $G_0 \in L^\infty(\mathbb{T}^n)$, such that*

$$\widehat{\phi}(\mathbf{t})\left([\widehat{\phi},\widehat{\phi}](\mathbf{t})\right)^{-1/2} = G_0((\mathbf{x})\widehat{\phi}(\mathbf{x})\left([\widehat{\phi},\widehat{\phi}]((\mathbf{x})\right)^{-1/2} \quad \textit{a.e. on}\, \mathbb{R}^n, \tag{25}$$

*where $\mathbf{x} = (A^*)^{-1}\mathbf{t}$;*
*($\mathbf{C}_3$) The origin is a point of $(G, A^*)$-approximate continuity of the function $|\widehat{\phi}|^2 \cdot [\widehat{\phi}, \widehat{\phi}]^{-1}$ if we set $|\widehat{\phi}(\mathbf{0})|^2([\widehat{\phi}, \widehat{\phi}](\mathbf{0}))^{-1} = 1$, and there exists $G_0 \in L^\infty(\mathbb{T}^n)$, a $\mathbb{Z}^n$-periodic function such that (25) holds.*

## 4. On the Existence of $H^2_G$-MRA and $H^2_G$-FMRA

In this section we discuss the following question:
Let $A : \mathbb{R}^n \to \mathbb{R}^n$ be a linear invertible map such that the condition (*) holds. Let $G \subseteq \mathbb{R}^n$ be a measurable $A^*$-set, $|G|_n > 0$. Do there exist $H^2_G$-MRA or $H^2_G$-FMRA associated with $A$?

If we look for an $H^2_G$-MRA associated with $A$ for which the scaling function $\phi \in H^2_G$ is such that $\widehat{\phi} = \chi_E$, where $E \subset G$ is a measurable set of positive Lebesgue measure then by Theorem 3.1 we obtain the following characterization of such sets.

**Theorem 4.1.** *Let $\phi \in H^2_G$ be such that $|\widehat{\phi}| = \chi_E$, where $E \subset G$ is a measurable set of positive Lebesgue measure. Then $\phi$ is a scaling function for an $H^2_G$-MRA associated with A if and only if following conditions hold:*

*1) $E \subset G$, and $dn_{A^*}(E) = dn_{A^*}(G)$;*
*2) $|(E+\mathbf{k}) \bigcap (E+\mathbf{m})|_n = 0 \quad$ if $\quad \mathbf{k} \neq \mathbf{m} \quad$ and $\quad \mathbf{k}, \mathbf{m} \in \mathbb{N}^n$;*
*3) $\left|\mathbb{R}^n \setminus (\bigcup_{\mathbf{k}\in\mathbb{Z}^n}(E+\mathbf{k}))\right|_n = 0$;*
*4) $E \subseteq A^*E$.*

In Theorem 4.1 and further in this section the notation $E \subset \Omega$, where $E$ and $\Omega$ are sets in $\mathbb{R}^n$ means that $|E \setminus \Omega|_n = 0$.

A characterization of scaling functions $\phi$ in an $A-$MRA, such that $|\widehat{\phi}|$ is a characteristic function of a measurable set was given by W. R. Madych [30]. Independently, M. Papadakis [34] has given another characterization when such an MRA is defined on $L^2(\mathbb{R})$. Afterwards, the later result was improved by M. Papadakis, H. Sikić, G. Weiss [35] and extended for the general case by M. Bownik, Z. Rzeszotnik and D. Speegle [5].

**Proof of Theorem 4.1.** If a function $\phi \in H^2_G$ such that $|\widehat{\phi}| = \chi_E$ is a scaling function for an $H^2_G$-MRA associated with the map $A$, which satisfies to the condition (*), then the condition ($C_2$) of Theorem 3.1 holds. The inclusion $E \subset G$ follows from the hypothesis $\phi \in H^2_G$. Hence, knowing that the origin is a point of $(G, A^*)$-continuity of the function $|\widehat{\phi}|$ if $|\phi(0)| = 1$ we obtain that $\mathrm{dn}_{A^*}(E) = \mathrm{dn}_{A^*}(G)$. Conditions 2) and 3) follow from Lemma 3.4. Moreover, it is well known that if conditions 2) and 3) hold then $|E|_n = 1$ (see [30], [39]). The condition 4) follows immediately from the condition (c) of Theorem 3.1.
If $E \subset G$ is a measurable set of positive Lebesgue measure such that the conditions 1)–4) hold, then it is easy to check that the conditions (a) and (b) are true. Let us check that the condition (c) also holds. We set

$$H(\mathbf{t}) = \sum_{\mathbf{k}\in\mathbb{Z}^n} \frac{\widehat{\phi}(A^*(\mathbf{t}+\mathbf{k}))}{\widehat{\phi}(\mathbf{t}+\mathbf{k})} \quad \text{for} \quad \mathbf{t} \in \mathbb{R}^n. \tag{26}$$

By conditions 2) and 4) we can find a set $R_0 \subset \mathbb{R}^n$, $|R_0|_n = 0$ such that for any $\mathbf{t} \in \mathbb{R}^n \setminus R_0$ the series in (26) has at most one non vanishing term. Hence, the function $H$ is well defined on $\mathbb{R}^n \setminus R_0$ and the modulus of $H$ on that set is one or zero. Evidently, the function $H$ is $\mathbb{Z}^n$-periodic and satisfies to the equation (24). $\square$

If we look for $\phi \in H^2_G$ such that $|\widehat{\phi}| = \chi_E$, where $E \subset G$ is a measurable set of positive Lebesgue measure and $\phi$ is a scaling function of an $H^2_G$-FMRA then, by Theorem 3.1, we obtain the following characterization.

**Theorem 4.2.** *Let $\phi \in H^2_G$ be such that $|\widehat{\phi}| = \chi_E$, where $E \subset G$ is a measurable set of positive Lebesgue measure. Then $\phi$ is a scaling function for an $H^2_G$-FMRA associated with A if and only if following conditions hold:*

*1*) $E \subset G$, and $dn_{A^*}(E) = dn_{A^*}(G)$;*
*2*) $\sum_{\mathbf{k}\in\mathbb{Z}^n} \chi_E(\cdot + \mathbf{k}) \in L^\infty(\mathbb{R}^n)$;*
*4*) $E \subseteq A^*E$.*

**Proof.** Comparing the formulations of Theorem 3.1 and Corollary 3.1 one concludes that the differences are connected with Lemma 3.4 and Theorem L. It is evident that the condition (14) holds for a function $|\widehat{\phi}| = \chi_E$ if and only if $\sum_{\mathbf{k}\in\mathbb{Z}^n} \chi_E(\cdot + \mathbf{k}) \in L^\infty(\mathbb{R}^n)$. The rest of the proof is similar to the proof of Theorem 4.1. $\square$

Improving relevant results in [16] and [14] Q. Gu and D. Han [22] proved the following result.

**Theorem M.** *Let $A : \mathbb{R}^n \to \mathbb{R}^n$ be a linear invertible map such that the condition (*) holds. Let $G \subseteq \mathbb{R}^n$ be a measurable $A^*$-set, $|G|_n > 0$, and $measG_{\mathbb{T}^n} = 1$. Then there exists $\phi \in H^2_{G}$ such that $\widehat{\phi} = \chi_E$, where $E \subset G$, which is a scaling function of an $H^2_G$-MRA.*

The method used in [22] can be slightly modified to yield the following result.

**Lemma N.** *Let $A : \mathbb{R}^n \to \mathbb{R}^n$ be a linear invertible map such that the condition (*) holds. Let $G \subseteq \mathbb{R}^n$ be a measurable $A^*$-set, $|G|_n > 0$. Then there exists $E \subset G$ such that $G \bigcap B_r \subset E$ for some $r > 0$, $meas(G_{\mathbb{T}^n} \setminus E_{\mathbb{T}^n}) = 0$ and the conditions 2) and 4) hold.*

The following Lemma is contained in the proof of the main result of the article by X. Dai, Y. Diao, Q. Gu, D. Han [12].

**Lemma O.** *Let $A : \mathbb{R}^n \to \mathbb{R}^n$ be a linear invertible expansive map. If $G \subseteq \mathbb{R}^n$ is a measurable $A^*$-set, $|G|_n > 0$ then $meas\, G_{\mathbb{T}^n} = 1$.*

By Lemmas N and O we have

**Theorem P.** *Let $A : \mathbb{R}^n \to \mathbb{R}^n$ be a linear invertible map such that the condition (*) holds. Let $G \subseteq \mathbb{R}^n$ be a measurable $A^*$-set, $|G|_n > 0$. Then there exists $\phi \in H^2_G$ such that $\widehat{\phi} = \chi_E$, where $E \subset G$ which is a scaling function of an $H^2_G$-MRA.*

## References


[1] J. J. Benedetto, S. Li; *The theory of multiresolution analysis frames and applications to filter banks* , Appl. Comput. Harmon. Anal. 5 (1998), no. 4, 389–427.

[2] J. J. Benedetto, D. Walnut; *Gabor frames for $L^2$ and related spaces*, in *Wavelets: Mathematics and Applications,* Eds. J. J. Benedetto, M. W. Frazier (CRC Press, Boca Raton 1994).

[3] C. de Boor, R.A. DeVore, A. Ron; *The structure of finitely generated shift-invariant spaces in $L_2(\mathbb{R}^d)$*, J. Funct. Anal. 119 (1994), no. 1, 37–78.

[4] C. de Boor, R. DeVore, A. Ron; *On the construction of multivariate (pre)wavelets,* Constr. Approx. 9 (1993), 123–166.

[5] M. Bownik, Z. Rzeszotnik, D. Speegle; *A characterization of dimension functions of wavelets,* Appl. Comput. Harmon. Anal. 10 (2001), no. 1, 71–92.

[6] A. Bruckner; *Differentiation of real functions* , Lecture Notes in Mathematics, 659. Springer, Berlin, 1978.

[7] A. Calogero; *Wavelets on general lattices, associated with general expanding maps of* $\mathbb{R}^n$, Electron. Res. Announc. Amer. Math. Soc. 5 (1999), 1–10.

[8] Charles K. Chui; *An Introduction to Wavelets,* Academic Press, Inc. 1992.

[9] O. Christensen; *An Introduction to frames and Riesz bases,* Birkhäser, Boston, 2003.

[10] P. Cifuentes, K.S. Kazarian, A. San Antolín; *Characterization of scaling functions,* Wavelets and splines: Athens 2005, 152–163, Mod. Methods Math., Nashboro Press, Brentwood, TN, 2006.

[11] P. Cifuentes, K. S. Kazarian, A. San Antolín; *Characterization of scaling functions in a Multiresolution Analysis,* Proc. of Amer. Math. Soc., 133 (2005), 1013–1022.

[12] X. Dai, Y. Diao, Q. Gu, D. Han; *The existence of subspace wavelet sets,*. Approximation theory, wavelets and numerical analysis (Chattanooga, TN, 2001). J. Comput. Appl. Math. 155 (2003), no. 1, 83–90.

[13] X. Dai, Y. Diao, Q. Gu, D. Han; *Frame wavelets in subspaces of $L^2(\mathbb{R}^d)$*, Proc. Amer. Math. Soc. 130 (2002), no. 11, 3259–3267.

[14] X. Dai, D. R. Larson, D. M. Speegle; *Wavelet sets in $\mathbb{R}^n$. II.* Wavelets, multiwavelets, and their applications (San Diego, CA, 1997), 15–40, Contemp. Math., 216, Amer. Math. Soc., Providence, RI, 1998.

[15] X. Dai, D. Larson, D. M. Speegle; *Wavelet sets in $\mathbb{R}^n$*, J. Fourier Anal. Appl., 3 (1997), no. 4, 451-456.

[16] X. Dai, S. Lu; *Wavelets in subspaces,* Michigan Math. J. 43 (1996), no. 1, 81–98.

[17] I. Daubechies; *Ten lectures on wavelets,* SIAM, Philadelphia, 1992.

[18] I. Daubechies, B. Han, A. Ron, Z. Shen; *Framelets: MRA-based constructions of wavelet frames,* Appl. Comput. Harmon. Anal., 14 (2003), 1–46.

[19] V. Dobrić, R. F. Gundy, P. Hitczenko; *Characterizations of orthonormal scale functions: a probabilistic approach,* J. Geom. Anal. 10 (2000), no. 3, 417–434.

[20] R. Duffin, A. Schaeffer; *A class of nonharmonic Fourier series,* Trans. of Amer. Math. Soc., 72 (1952), 341–366.

[21] K. Gröchening, W. R. Madych; *Multiresolution analysis, Haar bases and self-similar tillings of* $\mathbb{R}^n$ , IEEE Trans. Inform. Theory, 38(2) (March 1992), 556–568.

[22] Q. Gu, D. Han; *On Multiresolution Analysis (MRA) Wavelets in $\mathbb{R}^n$,* J. of Fourier Anal. Appl., 6, n 4 (2000), 437–447.

[23] D. Han, D. R. Larson; *Frames, bases and group representation* , Memoirs of the AMS, 697, Providence, RI, 2000.

[24] H. Helson; *Harmonic Analysis* , The Wadsworth & Brooks/ Cole mathematics series, 1983.

[25] E. Hernández, G. Weiss; *A first course on Wavelets*, CRC Press, Inc. 1996.

[26] E. Hernández, X. Wang, G. Weiss; *Characterization of wavelets, scaling functions and wavelets associated with multiresolution analyses*, Function spaces, interpolation spaces, and related topics (Haifa, 1995), 51–87, Israel Math. Conf. Proc., 13, Bar-Ilan Univ., Ramat Gan, 1999.

[27] R. Q. Jia, C. A. Micchelli; *Using the refinement equations for the construction of pre-wavelets. II. Powers of two. Curves and surfaces*, 209–246, Academic Press, Boston, MA, 1991.

[28] H. O. Kim, R. Y. Kim, J. K. Lim; *On the spectrums of frame multiresolution analyses*, J. Math. Anal. Appl. 305 (2005), no. 2, 528–545.

[29] R. A. Lorentz, W. R. Madych, A. Sahakian; *Translation and dilation invariant subspaces of $L^2(\mathbb{R})$ and multiresolution analyses,* Applied and Computational Harmonic Analysis **5** (1998), no. 4, 375–388.

[30] W. R. Madych; *Some elementary properties of multiresolution analyses of $L^2(\mathbb{R}^d)$*, Wavelets - a tutorial in theory and applications, Ch. Chui ed., Wavelet Anal. Appl. 2, Academic Press (1992), 259–294.

[31] S. Mallat; *Multiresolution approximations and wavelet orthonormal bases for $L^2(R)$*, Trans. of Amer. Math. Soc., 315 (1989), 69–87.

[32] Y. Meyer; *Ondelettes et opérateurs. I*, Hermann, Paris (1996) [ English Translation: Wavelets and operators, Cambridge University Press, (1992).]

[33] I. P. Natanson; *Theory of functions of a real variable* London, vol. II, 1960.

[34] M. Papadakis; *Unitary mappings between multiresolution analyses of $L^2(\mathbb{R}^n)$ and a parameterization of low pass filters,* J. of Fourier Analysis and Applications, Vol. 4 (1998), pp. 199–214.

[35] M. Papadakis, H. Sikić, G. Weiss; *The characterization of low pass filters and some basic properties of wavelets, scaling functions and related concepts*, J. Fourier Anal. Appl. 5 (1999), no. 5, 495–521.

[36] A. Ron, Z. Shen; *Frames and stable bases for shift invariant subspaces of $L^2(\mathbb{R}^n)$*, Can. J. Math. 47, no. 5. (1995), 1051–1094.

[37] A. Ron, Z. Shen; *Affine systems in $L_2(\mathbb{R}^d)$ : The analysis of the analysis operator,* J. Funct. Anal., 148 (1997), 408–447.

[38] R. Strichartz; *Construction of orthonormal wavelets*, Wavelets: mathematics and applications, 23–50, Stud. Adv. Math., CRC, Boca Raton, FL, 1994.

[39] P. Wojtaszczyk; *A mathematical introduction to wavelets* London Mathematical Society, Student Texts 37 (1997).

# AN ABSTRACT COIFMAN–ROCHBERG–WEISS COMMUTATOR THEOREM

JOAQUIM MARTIN*

*Department of Mathematics, Universidad Autònoma de Barcelona, 08193 Bellaterra (Barcelona) Spain*
*E-mail:jmartin@mat.uab.es*

MARIO MILMAN

*Department of Mathematics, Florida Atlantic University Boca Raton, Florida 33431*
*E-mail:extrapol@bellsouth.net, http://www.math.fau.edu/milman*

We formulate and prove a version of the celebrated Coifman–Rochberg–Weiss commutator theorem for the real method of interpolation

*Keywords*: Commutators, interpolation spaces, BMO.

## Dedication

*It is a special pleasure for us to dedicate this paper to you, our dear friend Dan Waterman on the occasion of your 80th birthday. But that is not all. The things we discuss here are intimately related to important work by another dear friend, and so to you too Richard Rochberg, warmest greetings on the occasion of your 65th birthday. We wish both of you many many more wonderful and creative years.*

## 1. Introduction

Commutator estimates play an important role in analysis (cf. [20]). Our starting point in this paper is the celebrated commutator theorem of Coifman–Rochberg–Weiss [5]. Let $K$ be a Calderón–Zygmund operator, and let $b \in BMO(R^n)$. Denote by $M_b$ the operator "multiplication by $b$", then (cf. [5])

$$\left\|[K, M_b]f\right\|_p \leq c \left\|b\right\|_{BMO} \left\|f\right\|_p, 1 < p < \infty, \tag{1}$$

where $[K, M_b]f = K(bf) - bK(f)$. Since each of the operators $f \to K(bf)$ and $f \to bK(f)$ is unbounded on $L^p$, the cancellation that results of taking their difference is essential for the validity of (1).

The Coifman–Rochberg–Weiss commutator theorem has found many applications in the study of PDEs, Jacobians, Harmonic Analysis, and was also the starting point of the Rochberg–Weiss [19] abstract theory of commutators in the

*The first author is supported in part by MTM2007-60500 and by CURE 2005SGR 00556.

setting of scales of interpolation spaces, which itself has had many applications (cf. [12], [13], [18], and the references therein).

It is instructive to review informally one of the proofs of (1) provided in [5]. Suppose that $b \in BMO$, and fix $p > 1$. Then it is well known that we can find $\varepsilon > 0$ small enough such that, for all $0 < \alpha < \varepsilon$, $e^{\alpha b}$ and $e^{-\alpha b} \in A_p$ (here $A_p$ is the class of Muckenhoupt weights). Let $K$ be a CZ operator, then $K$ is bounded on the weighted spaces $L^p(e^{\alpha b}), |\alpha| < \varepsilon$. In other words, the family of operators $f \to e^{\alpha b}K(e^{-\alpha b}f)$ is uniformly bounded on $L^p$ for $|\alpha| < \varepsilon$. It follows readily that one can extended these operators to an analytic family of operators $T(z)f = e^{zb}K(e^{-zb}f)$, for $|z| < \varepsilon$, and then show that, $\frac{d}{dz}T(z)f\big|_{z=0} = \frac{1}{2}[K, M_b]f$ is also a bounded operator on $L^p$. In particular, it follows that, in the statement of the theorem, we can replace CZ operators by operators $T$ with the same weighted norm inequalities, i.e. the result holds for any operator $T$, such that for all weights in the $A_p$ class of Muckenhoupt, $T : L^p(w) \to L^p(w), 1 < p < \infty$, boundedly.

The previous argument was the starting point of the Rochberg–Weiss [19] theory of abstract commutator estimates for the complex method of interpolation, later extended to the real method by these authors jointly with Jawerth (cf. [11]). The subject has been intensively developed in the last 30 years (cf. the recent survey by Rochberg [18] and the references therein).

While the Rochberg–Weiss theory, when suitably specialized to weighted $L^p$ spaces, can be used to re-prove the Coifman–Rochberg–Weiss commutator theorem, in this paper we consider a different problem: we give an abstract formulation of the Coifman–Rochberg–Weiss commutator theorem which is valid for interpolation scales themselves. Since we work with the real method, the cancellations will be exploited via integration by parts and a suitable re-interpretation of the relevant $BMO$ condition.

Before we formulate our main result let us recall some basic definitions associated with the real method of interpolation (cf. [3] for more details). Let $\bar{X} = (X_0, X_1)$ be a compatible pair of Banach spaces. To define the real interpolation spaces[a] $(X_0, X_1)_{\theta,q}$ we start by considering on $X_0 \cap X_1$ the family of norms

$$J(t, x; \bar{X}) = \max\{\|x\|_{X_0}, t\|x\|_{X_1}\}, t > 0.$$

Let $\theta \in (0, 1), 1 \leq q \leq \infty$. We consider the elements $f \in X_0 + X_1$, that can be represented by

$$f = \int_0^\infty u(s)\frac{ds}{s} \text{ (crucially here the convergence of the integral is in the } X_0{+}X_1 \text{ sense)},$$

where $u : (0, \infty) \to X_0 \cap X_1$. We let

$$\Phi_{\theta,q}(g) = \left\{\int_0^\infty \left(s^{-\theta}|g(s)|\right)^q \frac{ds}{s}\right\}^{1/q},$$

[a] We shall only consider the $J-$method in this note.

$$\bar{X}_{\theta,q} = \{f = \int_0^\infty u(s)\frac{ds}{s} \text{ in } X_0 + X_1 : \Phi_{\theta,q}(J(s,u(s);\bar{X})) < \infty\},$$

$$\|f\|_{\bar{X}_{\theta,q}} = \inf\{\Phi_{\theta,q}(J(s,u(s);\bar{X})) : f = \int_0^\infty u(s)\frac{ds}{s} \text{ in } X_0 + X_1\}.$$

Likewise, if $w$ is a positive function on $(0,\infty)$, we define the corresponding spaces $\bar{X}_{\theta,q,w}$ by means of the use of the function norm

$$\Phi_{\theta,q,w}(g) = \Phi_{\theta,q}(wg).$$

In this setting we consider the nonlinear operator

$$f \to u_f : (0,\infty) \to X_0 \cap X_1,$$

where $u_f$ has been selected so that

$$f = \int_0^\infty u_f(s)\frac{ds}{s} \text{ in } X_0 + X_1, \tag{2}$$

and[b]

$$\Phi_{\theta,q}(J(s,u_f(s);\bar{X})) \leq 2\|f\|_{\bar{X}_{\theta,q}}.$$

We then define

$$\Omega f = \Omega_{\bar{X}} f = \int_0^\infty u_f(s)\log s\frac{ds}{s}. \tag{3}$$

The commutator theorem in this context (cf. [11]) states that if $T : \bar{X} \to \bar{Y}$ is a bounded linear operator, then the nonlinear operator

$$\begin{aligned}[T,\Omega]f &= T(\Omega_{\bar{X}} f) - \Omega_{\bar{Y}}(Tf) \\ &= \int_0^\infty (T(u_f(s)) - u_{Tf}(s))\log s\frac{ds}{s}\end{aligned} \tag{4}$$

is bounded,

$$\|[T,\Omega]f\|_{\bar{Y}_{\theta,q}} \leq c\|T\|_{\bar{X}\to\bar{Y}}\|f\|_{\bar{X}_{\theta,q}}.$$

One possible interpretation of the appearance of the logarithm in formula (3) (and hence (4)) can be given if we try to imitate the arguments of Coifman–Rochberg–Weiss and bring into the argument analytic functions with suitable cancellations. Indeed, if we represent the elements of $\bar{X}_{\theta_0,q}$ using the normalization $u_{\theta_0 f}(s) = s^{\theta_0}u_f(s)$, then the elements in $\bar{X}_{\theta_0,q}$ can be represented by analytic functions (with appropriate control),

$$F(z) = \int_0^\infty s^{(z-\theta_0)}(u_{\theta_0 f}(s))\frac{ds}{s},\ F(\theta_0) = f.$$

In this setting we have

$$F'(\theta_0) = \Omega f.$$

[b]We use 2 for definiteness, obviously can replace 2 by $1+\varepsilon$.

The crucial point of the cancellation argument is that, while operators represented by derivatives of analytic functions can be unbounded (since we may lose control of the norm estimates), the canonical representation of $[T,\Omega]$

$$G'(\theta_0) = [T,\Omega]\, f,$$

with

$$G(z) = \int_0^\infty s^{(z-\theta_0)}(Tu_{\theta_0 f}(s) - u_{\theta_0 Tf}(s))\frac{ds}{s},$$

exhibits the crucial cancellation

$$\begin{aligned} G(\theta_0) &= \int_0^\infty (Tu_{\theta_0 f}(s) - u_{\theta_0 Tf}(s))\frac{ds}{s} \\ &= Tf - Tf \\ &= 0, \end{aligned} \tag{5}$$

which allows us to control the norm of $G'(\theta_0)$.

It is, of course, possible to eliminate all references to analytic functions, and formulate the results in terms of representations that exhibit cancellations. From this point of view the "badness" of the commutators is expressed by the fact that their canonical representations have an extra unbounded log factor (cf. (4)) which would lead to the weaker estimate

$$[T,\Omega] : \bar{X}_{\theta,q} \to \bar{Y}_{\theta,q,\frac{1}{(1+|\log s|)}}, \text{ (note that } \bar{Y}_{\theta,q} \subsetneqq \bar{Y}_{\theta,q,\frac{1}{(1+|\log s|)}}).$$

Here is where the cancellation (5), now expressed without reference to analytic functions, simply as an integral equal to zero, comes to our rescue and allows us to integrate by parts to find the "better" representation,

$$[T,\Omega]\, f = \int_0^\infty \Big(\int_0^t (Tu_{\theta_0 f}(s) - u_{\theta_0 Tf}(s))\frac{ds}{s}\Big)\frac{ds}{s}, \tag{6}$$

which leads to the correct estimate

$$[T,\Omega] : \bar{X}_{\theta,q} \to \bar{Y}_{\theta,q}.$$

This point of view was developed in [15].

To formulate the Coifman–Rochberg–Weiss theorem in our setting we give a different interpretation to the logarithm that appears in the formulae. First, for a given weight $w$ we introduce the (possibly non linear) operators $\Omega_w$, defined by

$$\Omega_w(f) = \int_0^\infty u_f(s)w(s)\frac{ds}{s}.$$

It follows that for $w \in L^\infty(0,\infty)$, the corresponding $\Omega_w$ is (trivially) a bounded operator,

$$\|\Omega_w(f)\|_{\bar{X}_{\theta,q}} \le c\,\|w\|_{L^\infty}\,\|f\|_{\bar{X}_{\theta,q}},$$

and therefore the corresponding commutators $[T,\Omega_w]$ are also bounded. On the other hand, for the mildly unbounded function $w(s) = \log(s)$, we have $\Omega_w = \Omega$, which is not bounded on $\bar{X}_{\theta,q}$, but for which cancellations imply the boundedness of commutators of the form $[T,\Omega]$. Now, as is well known, the logarithm is a typical example of a function with $BMO$ behavior. Therefore we now ask more generally: for which weights $w$ can we assert that for all bounded linear operators $T:\bar{X}\to\bar{Y}$, we have that $[T,\Omega_w]$ is a bounded operator as well? The answer to this question is what we shall call "the abstract Coifman–Rochberg–Weiss theorem."

Not surprisingly the answer is given in terms of a suitable $BMO$ type space which allows us to control the oscillations of $w$. Let $Pw(t)=\frac{1}{t}\int_0^t w(s)ds$ and define

$$w^{\#}(t)=Pw(t)-w(t)=\frac{1}{t}\int_0^t w(s)ds-w(t)=\frac{1}{t}\int_0^t (w(s)-w(t))\,ds.$$

Then we consider the following analog[c] of $BMO(R_+)$ introduced in [16]:

$$W=\{w:w^{\#}(t)\in L^\infty(0,\infty)\},\quad \text{with}\quad \|w\|_W=\|Pw-w\|_{L^\infty}.$$

There is a direct connection between $W$ and the space $L(\infty,\infty)$ of Bennett–DeVore–Sharpley [4]:

$$w\in L(\infty,\infty)\Leftrightarrow w^*\in W,$$

where $w^*$ denotes the non-increasing rearrangement of $w$. In particular, we note that, as expected, the log has bounded oscillation since

$$(\log t)^{\#}=\frac{1}{t}\int_0^t \log s ds-\log t=-1.$$

It will turn out that $W$ is the correct way to measure oscillation in our context. In particular, we will show below that, when dealing with the commutators $[T,\Omega_w]$, the corresponding "good representation" (cf. (6)) is given by

$$[T,\Omega_w]f=\int_0^\infty \Big(\int_0^t (Tu_f(s)-u_{Tf}(s))\frac{ds}{s}\Big)w^{\#}(s)\frac{ds}{s}.$$

The purpose of this note is to prove the following abstract analog of the Coifman–Rochberg–Weiss commutator theorem

**Theorem 1.1.** *Suppose that $w\in W$, and let $\bar{X},\bar{Y}$, be Banach pairs. Then, for any bounded linear operator $T:\bar{X}\to\bar{Y}$, the commutator $[T,\Omega_w]$ is bounded, $[T,\Omega_w]:\bar{X}_{\theta,q}\to\bar{Y}_{\theta,q}$, $0<\theta<1, 1\le q\le\infty$, and, moreover,*

$$\|[T,\Omega_w]f\|_{\bar{Y}_{\theta,q}}\le c\|T\|_{\bar{X}\to\bar{Y}}\|w\|_W\|f\|_{\bar{X}_{\theta,q}}.$$

We will also prove higher order versions of this result (cf. [15] and the references therein). Using the strong form of the fundamental lemma (cf. [6]) one can connect the results above with those obtained in [1] for the $K-$method, and, moreover, give explicit instances of these operators.

[c]For martingales it can be explicitly shown, by means of selecting appropriate sigma fields (cf. [10]), that $W$ is a $BMO$ martingale space. $W$ has also appeared before in several papers on interpolation theory (cf. [9], [2]).

## 2. Representation Theorems

As we have indicated in the Introduction, commutator theorems can be formulated as results about special representations of certain elements in interpolation scales. To develop our program explicitly it will be necessary to integrate by parts often, so we start by collecting some elementary calculations that will be useful for that purpose.

**Lemma 2.1.** *The operator $P$ is bounded on $W$.*

**Proof.**

$$\begin{aligned}(Pw)^{\#}(t) &= \frac{1}{t}\int_0^t Pw(s)ds - Pw(t) \\ &= \frac{1}{t}\int_0^t (Pw(s) - w(s))\,ds + Pw(t) - Pw(t).\end{aligned}$$

Therefore,

$$\left|(Pw)^{\#}(t)\right| \leq \|w\|_W \cdot \qquad \square$$

**Lemma 2.2.** *Let $w \in W$, and let $0 < \theta < 1$. Then*

$$\lim_{t\to 0} t^{\theta} w(t) = \lim_{t\to\infty} t^{-\theta} w(t) = 0.$$

**Proof.** *Write $Pw = w^{\#} + w$, then, since $w^{\#}$ is bounded, $\lim_{t\to 0} t^{\theta} w^{\#}(t) = \lim_{t\to\infty} t^{-\theta} w^{\#}(t) = 0$, and we see that it is enough to show that $\lim_{t\to 0} t^{\theta} Pw(t) = \lim_{t\to\infty} t^{-\theta} Pw(t) = 0$. Now, from $tPw(t) = \int_0^t w(s)ds$, we get $(Pw)'(t) = -\frac{Pw(t)-w(t)}{t}$. Therefore,*

$$\begin{aligned}|Pw(t)| &\leq |Pw(1)| + \left|\int_t^1 w^{\#}(s)\frac{ds}{s}\right| \\ &\leq \|w\|_W (1 + |\log t|).\end{aligned}$$

*and the result follows.* $\square$

Although we shall not make use of the next result in this section it is convenient to state it here to stress the $BMO$ characteristics of the space $W$.

**Lemma 2.3.** *(i) (cf. [1]) Let $\overline{Q}f(t) = \int_t^1 f(s)\frac{ds}{s}$ then*

$$W = L_\infty + \overline{Q}(L_\infty).$$

*(ii) Let $W_1 = \left\{w : \sup_s |sw'(s)| < \infty\right\}$. Then, $W_1 \subset W$.*

*(iii)*

$$W = L_\infty + W_1.$$

**Proof.** (i) see [1].

(ii) Suppose that $w \in W_1$. Integrating by parts

$$\frac{1}{x}\int_0^x sw'(s)ds = \frac{1}{x}\, sw(s)|_{s=0}^{s=x} - \frac{1}{x}\int_0^x w(s)ds.$$

It is easy to see (cf. Lemma 2.2) that $\lim_{x\to 0} sw(s) = 0$, hence

$$\left|w^{\#}(x)\right| = \left|P(sw'(s))(x)\right|.$$

Consequently, since $P$ is bounded on $L^\infty$, it follows that $w^{\#} \in L^\infty$ and therefore $w \in W$.

(iii) Suppose that $w \in W$. Since $\left|t\,(Pw)'\right| = \left|w^{\#}(t)\right|$, it follows that $Pw \in W_1$. The desired decomposition is therefore given by

$$w = \underbrace{(w - Pw)}_{L^\infty} + \underbrace{Pw}_{W_1}.$$

□

The next result gives the representation theorem that we need to prove Theorem 1.1.

**Theorem 2.1.** *Let $\overline{H} = (H_0, H_1)$ be a Banach pair, and suppose that $w \in W$. Suppose that an element $f \in H_0 + H_1$ can be represented as*

$$f = \int_0^\infty u(s)w(s)\frac{ds}{s},$$

*with*

$$\int_0^\infty u(s)\frac{ds}{s} = 0,\ \ \Phi_{\theta,q}(J(t,u(t);\overline{H})) < \infty.$$

*Then,*

$$f \in \overline{H}_{\theta,q},$$

*and, moreover,*

$$\|f\|_{\overline{H}_{\theta,q}} \leq c_{\theta,q}\, \|w\|_W\, \Phi_{\theta,q}(J(t,u(t);\overline{H})).$$

**Proof.** Write

$$\begin{aligned} f &= \int_0^\infty u(s)w(s)\frac{ds}{s} \\ &= \int_0^\infty u(s)(w(s) - Pw(s))\frac{ds}{s} + \int_0^\infty u(s)Pw(s)\frac{ds}{s} \\ &= I_1 + I_2. \end{aligned}$$

It is plain that

$$\|I_1\|_{\overline{H}_{\theta,q}} \leq \|w\|_W\, \Phi_{\theta,q}(J(t,u(t);\overline{H})).$$

It remains to estimate $I_2$. We integrate by parts:

$$I_2 = Pw(t)\int_0^t u(s)\frac{ds}{s}\Big]_0^\infty - \int_0^\infty \left(\int_0^t u(s)\frac{ds}{s}\right)(w(t) - Pw(t))\frac{dt}{t}.$$

The integrated term vanishes. Suppose first that $q > 1$. We can write

$$\begin{aligned}\left\|\int_0^t u(s)\frac{ds}{s}\right\|_{H_0} &\le \int_0^t J(s,u(s))\frac{ds}{s} \le \left(\int_0^t \left(\frac{J(s,u(s))}{s^\theta}\right)^q \frac{ds}{s}\right)^{1/q}\left(\int_0^t s^{\theta q'}\frac{ds}{s}\right)^{1/q'} \\ &\le c_{\theta,q}\Phi_{\theta,q}(J(t,u(t);\overline{H}))t^\theta.\end{aligned} \tag{7}$$

By Lemma 2.1 $Pw \in W$ and therefore we may apply Lemma 2.2 to conclude that

$$\begin{aligned}\lim_{t\to 0}|Pw(t)|\left\|\int_0^t u(s)\frac{ds}{s}\right\|_{H_0} &\le c_{\theta,q}\lim_{t\to 0}\Phi_{\theta,q}(J(t,u(t);\overline{H}))t^\theta\,|Pw(t)| \\ &= 0.\end{aligned}$$

Likewise, using the cancellation condition

$$\int_0^t u(s)\frac{ds}{s} = -\int_t^\infty u(s)\frac{ds}{s}, \tag{8}$$

we have that

$$\left\|\int_t^\infty u(s)\frac{ds}{s}\right\|_{H_1} \le c\Phi_{\theta,q}(J(t,u(t);\overline{H}))t^{-\theta},$$

and once again we can apply Lemma 2.2 and find that

$$\lim_{t\to\infty}|Pw(t)|\left\|\int_t^\infty u(s)\frac{ds}{s}\right\|_{H_1} = 0.$$

The case $q = 1$ is simpler. For example, instead of using Holder's inequality in (7) we write

$$\int_0^t J(s,u(s))\frac{ds}{s} = \frac{t^\theta}{t^\theta}\int_0^t J(s,u(s))\frac{ds}{s} \le t^\theta \int_0^t \frac{J(s,u(s))}{s^\theta}\frac{ds}{s}.$$

It remains to estimate the $\overline{H}_{\theta,q}$ norm of $I_2 = \int_0^\infty \left(\int_0^t u(s)\frac{ds}{s}\right)(w(t) - Pw(t))\frac{dt}{t}$. By definition,

$$\|I_2\|_{\overline{H}_{\theta,q}} \le \Phi_{\theta,q}(J(t)), \tag{9}$$

where

$$\begin{aligned}J(t) &= J(t, \left(\int_0^t u(s)\frac{ds}{s}\right)(w(t) - Pw(t));\overline{H}) \\ &\le \|w\|_W\left(\left\|\int_0^t u(s)\frac{ds}{s}\right\|_{H_0} + t\left\|\int_0^t u(s)\frac{ds}{s}\right\|_{H_1}\right).\end{aligned}$$

The first term on the right hand side can be estimated directly by Minkowski's inequality

$$\left\| \int_0^t u(s) \frac{ds}{s} \right\|_{H_0} \leq \int_0^t J(s, u(s); \overline{H}) \frac{ds}{s},$$

while for the second we argue that, by (8),

$$\begin{aligned} t \left\| \int_0^t u(s) \frac{ds}{s} \right\|_{H_1} &= t \left\| \int_t^\infty u(s) \frac{ds}{s} \right\|_{H_1} \\ &\leq t \int_t^\infty J(s, u(s); \overline{H}) \frac{ds}{s^2}. \end{aligned}$$

Altogether, we arrive at

$$J(t) \leq \|w\|_W \left( \int_0^t J(s, u(s); \overline{H}) \frac{ds}{s} + t \int_t^\infty J(s, u(s); \overline{H}) \frac{ds}{s^2} \right).$$

Therefore, applying the $\Phi_{\theta,q}$ norm on both sides of the previous inequality and then using Hardy's inequalities to estimate the right hand side, we get

$$\Phi_{\theta,q}(J(t)) \leq c_{\theta,q} \|w\|_W \, \Phi_{\theta,q} \left( J(t, u(t); \overline{H}) \right). \tag{10}$$

Combining (10) and (9)

$$\|I_2\|_{\overline{H}_{\theta,q}} \leq c_{\theta,q} \|w\|_W \, \Phi_{\theta,q}(J(t)),$$

and collecting the estimates for $I_1$ and $I_2$ we finally obtain

$$\|f\|_{\overline{H}_{\theta,q}} \leq c_{\theta,q} \|w\|_W \, \Phi_{\theta,q}(J(t, u(t); \overline{H}))$$

as we wished to show. □

We are now ready for the proof of Theorem 1.1.

**Proof.** Suppose that $T$ is a given bounded linear operator $T : \bar{X} \to \bar{Y}$, and let $w \in W$. Let $\tilde{u}(t) = ((u_{Tf}(t) - T(u_f(t)))$. Then

$$[T, \Omega_w] \, f = \int_0^\infty \tilde{u}(t) w(t) \frac{dt}{t}$$

with

$$\Phi_{\theta,q}(J(t, \tilde{u}(t); \bar{Y}) \leq c \, \|T\|_{\bar{X} \to \bar{Y}} \, \|f\|_{\overline{X}_{\theta,q}}.$$

Since, moreover,

$$\int_0^\infty \tilde{u}(t) \frac{dt}{t} = 0,$$

we can apply theorem 2.1 to conclude that

$$\|[T, \Omega_w] \, f\|_{\overline{Y}_{\theta,q}} \leq c \, \|w\|_W \, \|T\|_{\bar{X} \to \bar{Y}} \, \|f\|_{\overline{X}_{\theta,q}},$$

as we wished to show. □

## 3. Higher Order Cancellations

We adapt the analysis of [15] to handle higher order cancellations. The corresponding higher order commutator theorems that follow will be stated and proved in the next section.

**Theorem 3.1.** *Let $\overline{H}$ be a Banach pair, and let $w \in W$. Suppose that $f$ admits a representation*

$$f = \int_0^\infty u(s) \left(Pw(s)\right)^2 \frac{ds}{s},$$

*with*

$$\int_0^\infty u(s) \frac{ds}{s} = 0, \quad \int_0^\infty u(s) Pw(s) \frac{ds}{s} = 0; \quad \Phi_{\theta,q}(J(t, u(t); \overline{H})) < \infty$$

*then,*

$$f \in \overline{H}_{\theta,q},$$

*and, moreover,*

$$\|f\|_{\overline{H}_{\theta,q}} \leq c \, \|w\|_W^2 \, \Phi_{\theta,q}(J(t, u(t); \overline{H})).$$

**Proof.** *We will integrate by parts repeatedly. We start writing*

$$f = \int_0^\infty u(t) \left(Pw(t)\right)^2 \frac{dt}{t} = \int_0^\infty Pw(t) d\left(\int_0^t u(s) Pw(s) \frac{ds}{s}\right).$$

*Then,*

$$f = Pw(t) \int_0^t u(s) Pw(s) \frac{ds}{s} \Big]_0^\infty - \int_0^\infty \left(\int_0^t u(s) Pw(s) \frac{ds}{s}\right) (w(t) - Pw(t)) \frac{dt}{t},$$

*we will show below that the integrated term vanishes, then*

$$f = -\int_0^\infty \left(\int_0^t u(s) Pw(s) \frac{ds}{s}\right) (w(t) - Pw(t)) \frac{dt}{t}. \tag{11}$$

*Now we consider the inner integral and integrate by parts*

$$\int_0^t u(s) Pw(s) \frac{ds}{s} = \int_0^t Pw(s) d\left(\int_0^s u(r) \frac{dr}{r}\right),$$

*using the fact that (cf. the proof of Theorem 2.1) $\lim_{t \to 0} Pw(t) \int_0^t u(s) \frac{ds}{s} = 0$, we get*

$$\int_0^t u(s) Pw(s) \frac{ds}{s} = Pw(t) \int_0^t u(s) \frac{ds}{s} - \int_0^t \left(\int_0^r u(s) \frac{ds}{s}\right) (w(r) - Pw(r)) \frac{dr}{r}.$$

*Inserting this result back in (11) we find that*

$$\begin{aligned} f &= -\int_0^\infty \left(\int_0^t u(s) Pw(s) \frac{ds}{s}\right) (w(t) - Pw(t)) \frac{dt}{t} \\ &= \int_0^\infty \left(Pw(t) \int_0^t u(s) \frac{ds}{s}\right) w^{\#}(t) \frac{dt}{t} + \int_0^\infty \left(\int_0^t \left(\int_0^r u(s) \frac{ds}{s}\right) w^{\#}(r) \frac{dr}{r}\right) w^{\#}(t) \frac{dt}{t} \\ &= I_0 + I_1. \end{aligned}$$

*Integrating by parts $I_0$ we get*

$$I_0 = Pw(t)\int_0^t \left(w^{\#}(r)\int_0^r u(s)\frac{ds}{s}\right)\frac{dr}{r}\Big|_0^\infty + \int_0^\infty \left(\int_0^t \left(\int_0^r u(s)\frac{ds}{s}\right) w^{\#}(r)\frac{dr}{r}\right) w^{\#}(t)\frac{dt}{t},$$

*where once again the integrated term vanishes. Hence,*

$$I_0 = I_1.$$

*Therefore, if we let* $U(t) = 2\left(\int_0^t \left(\int_0^r u(s)\frac{ds}{s}\right) w^{\#}(r)\frac{dr}{r}\right) w^{\#}(t)$, $f$ *can be represented by*

$$f = \int_0^\infty U(t)\frac{dt}{t}.$$

*Now we estimate the corresponding $J-$functional,* $J(t) = J(t, U(t); \bar{H})$, *by*

$$2\|w\|_W \left(\left\|\int_0^t \left(\int_0^r u(s)\frac{ds}{s}\right) w^{\#}(r)\frac{dr}{r}\right\|_{H_0} + t\left\|\int_0^t \left(\int_0^r u(s)\frac{ds}{s}\right) w^{\#}(r)\frac{dr}{r}\right\|_{H_1}\right)$$
$$= 2\|w\|_W (C_0 + tC_1).$$

*We readily see that $C_0$ is majorized by*

$$C_0 \le \|w\|_W \int_0^t \left(\int_0^r J(s, u(s); \overline{H})\frac{ds}{s}\right)\frac{dr}{r} = \|w\|_W \int_0^t J(r, u(r); \overline{H}) \ln\frac{t}{r}\frac{dr}{r}.$$

*To handle $C_1$ we work with the integral inside the norm $H_1$ by first using* $\int_0^r u(s)\frac{ds}{s} = -\int_r^\infty u(s)\frac{ds}{s}$ *and then changing the order of integration. We find that*

$$C_1 = \left\|\lim_{\alpha\to 0} C(\alpha)\right\|_{H_1},$$

*where* $C(\alpha) = \int_0^t \int_\alpha^s w^{\#}(r)\frac{dr}{r}u(s)\frac{ds}{s} + \int_t^\infty \int_\alpha^t w^{\#}(r)\frac{dr}{r}u(s)\frac{ds}{s}$. *We compute $C(\alpha)$ using the formula* $(Pw)'(t) = -\frac{w^{\#}(t)}{t}$, *and we get*

$$C(\alpha) = Pw(\alpha)\int_0^t u(s)\frac{ds}{s} - \int_0^t Pw(s)u(s)\frac{ds}{s} + Pw(\alpha)\int_t^\infty u(s)\frac{ds}{s} - Pw(t)\int_t^\infty u(s)\frac{ds}{s}.$$

*Now by the cancellation conditions:*

$$\int_0^\infty u(s)\frac{ds}{s} = 0 \Longrightarrow Pw(\alpha)\int_0^t u(s)\frac{ds}{s} = -Pw(\alpha)\int_t^\infty u(s)\frac{ds}{s},$$

*and*

$$\int_0^\infty u(s)Pw(s)\frac{ds}{s} = 0 \Longrightarrow \int_0^t u(s)Pw(s)\frac{ds}{s} = -\int_t^\infty u(s)Pw(s)\frac{ds}{s},$$

*we have*

$$C(\alpha) = \int_t^\infty u(s)[Pw(s) - Pw(t)]\frac{ds}{s}$$
$$= \int_t^\infty u(s)\int_t^s w^{\#}(r)\frac{dr}{r}\frac{ds}{s}.$$

*All in all it follows that,*

$$C_1 \leq \|w\|_W \int_t^\infty \|u(s)\|_{H_1} \ln\frac{s}{t}\frac{ds}{s}$$
$$\leq \|w\|_W \int_t^\infty J(s,u(s);\overline{H}) \ln\frac{s}{t}\frac{ds}{s^2}.$$

*Summarizing,*

$$J(t) \leq 2\,\|w\|_W^2 \left(\int_0^t J(r,u(r);\overline{H}) \ln\frac{t}{r}\frac{dr}{r} + t\int_t^\infty J(r,u(r);\overline{H}) \ln\frac{r}{t}\frac{dr}{r^2}\right).$$

*Applying the $\Phi_{\theta,q}$ norm and Hardy's inequalities (twice) we finally obtain*

$$\|f\|_{\overline{H}_{\theta,q}} \leq c\Phi_{\theta,q}\left(J(t,u(t);\overline{H})\right)$$
$$\leq c\,\|w\|_W^2\,\Phi_{\theta,q}\left(J(t,u(t);\overline{H})\right).$$

*To conclude the proof it remains to verify that the integrated terms we have collected along the way effectively vanish. More precisely, it remains to prove that*

$$\lim_{t\to\xi} Pw(t)\int_0^t u(s)Pw(s)\frac{ds}{s} = 0, \quad \textit{for } \xi = 0,\infty, \tag{12}$$

*and*

$$\lim_{t\to\xi} Pw(t)\int_0^t \left((w(r) - Pw(r))\int_0^r u(s)\frac{ds}{s}\right)\frac{dr}{r} = 0, \textit{ for } \xi = 0,\infty. \tag{13}$$

*To handle these limits we shall assume that $q > 1$, the case $q = 1$ is easier (cf. the proof of Theorem 2.1 above). We start with (12):*

$$\left\|\int_0^t u(s)Pw(s)\frac{ds}{s}\right\|_{H_0} \leq \int_0^t J(s,u(s);\overline{H})\,|Pw(s)|\,\frac{ds}{s}$$
$$\leq \left(\int_0^t \left(\frac{J(s,u(s);\overline{H})}{s^\theta}\right)^q \frac{ds}{s}\right)^{1/q} \left(\int_0^t \left(s^\theta\,|Pw(s)|\right)^{q'} \frac{ds}{s}\right)^{1/q'}$$
$$\leq \left(\Phi_{\theta,q}(J(s,u(s);\overline{H}))\right) c\left(\int_0^t \left(s^\theta\,|w(s)|\right)^{q'} \frac{ds}{s}\right)^{1/q'},$$

*where the last inequality is obtained by Hardy's inequality. Let $\widetilde{\theta} > 0$ be such that $\theta - \widetilde{\theta} > 0$. Since $w \in W \Rightarrow Pw \in W$ (cf. Lemma 2.1), therefore, by Lemma 2.2, we have*

$$\left|t^{\widetilde{\theta}}Pw(t)\right| \leq 1 \quad \textit{(if t suff. close to 0).}$$

*Thus, for small t,*

$$\left(\int_0^t \left(s^\theta \left|Pw(s)\right|\right)^{q'} \frac{ds}{s}\right)^{1/q'} \leq \left(\int_0^t \left(s^{\theta-\tilde{\theta}}\right)^{q'} \frac{ds}{s}\right)^{1/q'} \leq ct^{\theta-\tilde{\theta}},$$

*and*

$$\lim_{t\to 0}\left\|Pw(t)\int_0^t u(s)w(s)\frac{ds}{s}\right\|_{H_0} \leq \lim_{t\to 0} ct^{\theta-\tilde{\theta}}\left|Pw(t)\right| = 0.$$

*The corresponding limit when $t \to \infty$ can be handled by the same argument if we first use the cancellation property $\int_0^t u(s)Pw(s)\frac{ds}{s} = -\int_t^\infty u(s)Pw(s)\frac{ds}{s}$ and then apply the $H_1$ norm.*

*To see (13) we note that*

$$\begin{aligned}
&\left|Pw(t)\right|\left\|\int_0^t (w(r) - Pw(r))\int_0^r u(s)\frac{ds}{s}\frac{dr}{r}\right\|_{H_0} \\
&\leq \|w\|_W \left|Pw(t)\right| \int_0^t J(s, u(s); \overline{H}) \ln\frac{t}{s}\frac{ds}{s} \\
&\leq \|w\|_W \left|Pw(t)\right| t^\theta \left(\Phi_{\theta,q}(J(s, u(s); \overline{H}))\right) t^{-\theta}\left(\int_0^t \left(s^\theta \ln\frac{t}{s}\right)^{q'}\frac{ds}{s}\right)^{1/q'}.
\end{aligned}$$

*Now, the term on the right hand side converges to zero when $t \to 0$ by Lemma 2.2 and the fact that near zero,*

$$t^{-\theta}\left(\int_0^t \left(s^\theta \ln\frac{t}{s}\right)^{q'}\frac{ds}{s}\right)^{1/q'} \leq t^{-\theta}\left(\int_0^t \left(s^\theta\frac{s}{t}\right)^{q'}\frac{ds}{s}\right)^{1/q'} \leq ct^{-\theta}t^{-1}t^{1+\theta}.$$

*Again the case $t \to \infty$ is reduced to the case $t \to 0$ by a familiar argument using cancellations.* □

**Corollary 3.1.** *Let $\overline{H}$ be a Banach pair, and let $w \in W$. Suppose that*

$$f = \int_0^\infty u(s)\left(w(s)\right)^2\frac{ds}{s},$$

*with*

$$\int_0^\infty u(s)\frac{ds}{s} = 0, \quad \int_0^\infty u(s)w(s)\frac{ds}{s} = 0, \quad \int_0^\infty u(s)Pw(s)\frac{ds}{s} = 0; \tag{14}$$

*and*

$$\Phi_{\theta,q}(J(t, u(t); \overline{H})) < \infty.$$

*Then,*

$$f \in \overline{H}_{\theta,q},$$

*and, moreover,*

$$\|f\|_{\overline{H}_{\theta,q}} \leq c\,\|w\|_W^2\,\Phi_{\theta,q}(J(t, u(t); \overline{H})).$$

**Proof.** Write

$$\int_0^\infty u(s)\,(w(s))^2\,\frac{ds}{s} = \int_0^\infty u(s)\,(w(s) - Pw(s))\,w(s)\frac{ds}{s} + \int_0^\infty u(s)w(s)Pw(s)\frac{ds}{s}.$$

Since

$$w(t)Pw(t) = \frac{(w(t))^2 + (Pw(t))^2 - (w(t) - Pw(t))^2}{2},$$

we have

$$\int_0^\infty u(s)\,(w(s))^2\,\frac{ds}{s} = 2\int_0^\infty u(s)\,(w(s) - Pw(s))\,w(s)\frac{ds}{s}$$
$$-\int_0^\infty u(s)\,(w(s) - Pw(s))^2\,\frac{ds}{s} + \int_0^\infty u(s)\,(Pw(s))^2\,\frac{ds}{s}.$$

We now show how to control each of these terms. Let $\tilde{u}(s) = u(s)(w(s) - Pw(s))$, by the cancellation conditions (14) it follows that $\int_0^\infty \tilde{u}(s)\frac{ds}{s} = 0$. Therefore we can apply Theorem 2.1 to conclude that $\int_0^\infty \tilde{u}(s)w(s)\frac{ds}{s} \in \overline{H}_{\theta,q}$. It follows that

$$\left\| 2\int_0^\infty u(s)\,(w(s) - Pw(s))\,w(s)\frac{ds}{s} \right\|_{\overline{H}_{\theta,q}} \leq c\,\|w\|_W^2\,\Phi_{\theta,q}(J(t,u(t);\overline{H})).$$

The second term is also under control since $(w(s) - Pw(s))^2$ is bounded. Finally we may apply Theorem 3.1 to control the remaining term. □

**Theorem 3.2.** *Let $\overline{H}$ be a Banach pair, and let $w_0, w_1 \in W$. Suppose that*

$$f = \int_0^\infty u(s)w_0(s)w_1(s)\frac{ds}{s},$$

*with*

$$\int_0^\infty u(s)\frac{ds}{s} = 0, \quad \int_0^\infty u(s)w_j(s)\frac{ds}{s} = 0, \quad \int_0^\infty u(s)Pw_j(s)\frac{ds}{s} = 0 \quad (j = 0, 1)$$

*and*

$$\Phi_{\theta,q}(J(t,u(t);\overline{H})) < \infty.$$

*Then,*

$$f \in \overline{H}_{\theta,q}$$

*and, moreover,*

$$\|f\|_{\overline{H}_{\theta,q}} \leq c\max\{\|w_0\|_W, \|w_1\|_W\}^2\Phi_{\theta,q}(J(t,u(t);\overline{H})).$$

**Proof.** Write

$$w_0(s)w_1(s) = \frac{(w_0(s) + w_1(s))^2 - w_0(s)^2 - w_1(s)^2}{2}$$

and apply Corollary 3.1. □

For $n > 2$ we proceed by induction and we obtain

**Theorem 3.3.** *Let $\overline{H}$ be a Banach pair, and let $w \in W$.*

*(i) Suppose that*

$$f = \int_0^\infty u(s)\,(Pw(s))^n \frac{ds}{s},$$

*with*

$$\int_0^\infty u(s)w(s)^k \frac{ds}{s} = 0, \ \int_0^\infty u(s)Pw(s)^k \frac{ds}{s} = 0, \quad (k = 0, \cdots, n-1);$$

*and*

$$\Phi_{\theta,q}(J(t,u(t);\overline{H})) < \infty.$$

*Then,*

$$f \in \overline{H}_{\theta,q},$$

*and, moreover,*

$$\|f\|_{\overline{H}_{\theta,q}} \leq c\Phi_{\theta,q}(J(t,u(t);\overline{H})).$$

*(ii) If*

$$f = \int_0^\infty u(s)\,(w(s))^n \frac{ds}{s},$$

*with*

$$\int_0^\infty u(s)w(s)^k \frac{ds}{s} = 0, \ \int_0^\infty u(s)Pw(s)^k \frac{ds}{s} = 0, \ \int_0^\infty u(s)w(s)^{n-k}Pw(s)^k \frac{ds}{s} = 0,$$

*$(k = 0, \cdots, n-1)$ and*

$$\Phi_{\theta,q}(J(t,u(t);\overline{H})) < \infty,$$

*then,*

$$f \in \overline{H}_{\theta,q},$$

*and, moreover,*

$$\|a\|_{\overline{H}_{\theta,q}} \leq c\,\|w\|_W^n\,\Phi_{\theta,q}(J(t,u(t);\overline{H})).$$

**Remark 3.1.** In the classical case (cf. [15], theorem 3) $w(t) = \ln t$, and therefore we have $Pw(t) = \ln t - 1$. Consequently the conditions

$$\int_0^\infty u(s)Pw(s)^k \frac{ds}{s} = 0, \int_0^\infty u(s)w(s)^{n-k}Pw(s)^k \frac{ds}{s} = 0, \ (k = 0, \cdots, n-1);$$

actually follow from

$$\int_0^\infty u(s)\,(w(s))^k \frac{ds}{s} = 0, \ (k = 0, \cdots, n-1).$$

## 4. Higher Order Commutators

We consider higher order commutators defined as follows (cf. [15], [1], [18]). Let $\bar{X}$ and $\bar{Y}$ be Banach pairs, and let $T : \bar{X} \to \bar{Y}$ be a bounded linear operator. Given a nearly optimal representation (cf. 2 above)

$$f = \int_0^\infty u_f(s)\frac{ds}{s}$$

we let

$$\Omega_{n,w}f = \frac{1}{n!}\int_0^\infty u_f(s)(w(s))^n\frac{ds}{s},\ n = 0, 1, ...$$

and form the commutators

$$C_{n,w}f = \begin{cases} Tf, & n = 0 \\ [T, \Omega_{1,w}]\, f, & n = 1 \\ [T, \Omega_{2,w}]\, f - \Omega_{1,w}(C_{1,w}f), & n = 2 \\ \cdots\cdots\cdots\cdots & \\ [T, \Omega_{n,w}]\, f - \Omega_{1,w}(C_{n-1,w}f) - \cdots \Omega_{n-1,w}(C_{1,w}f) & \end{cases}$$

Observe that the commutators $[T, \Omega_{n,w}]$ alone are not bounded and we need to form more complicated expressions like $C_{n,w}$ in order to produce the necessary cancellations. Moreover, since the operations $\Omega_{j,w}$ are not linear, simple minded iterations of the form $\Omega_{1,w}\,[T, \Omega_{1,w}] - [T, \Omega_{1,w}]\,\Omega_{1,w}$, *etc*, cannot be treated directly using Theorem 1.1.

**Theorem 4.1.** *Suppose that $w \in W$. Then the commutators $C_{n,w}$ are bounded, $C_{n,w} : \bar{X}_{\theta,q} \to \bar{Y}_{\theta,q}$, $0 < \theta < 1, 1 \leq q \leq \infty$, and, moreover, for each instance $g = w$, or $g = Pw$, we have*

$$\|C_{n,w}f\|_{\bar{Y}_{\theta,q}} \leq c\,\|T\|_{\bar{X}\to\bar{Y}}\,\|w\|_W^n\,\|f\|_{\bar{X}_{\theta,q}}\,.$$

**Proof.** *We only consider in detail the case $n = 2$. Writing $w = (w - Pw) + Pw$, we see that we only need to deal with the commutator $C_{2,Pw}$. Let*

$$u(s) = T(u_f(s)) - u_{T(f)}(s)$$

*then*

$$C_{2,Pw}(Tf) = \frac{1}{2}\int_0^\infty u(t)(Pw(t))^2\frac{dt}{t} - \int_0^\infty \widetilde{u}(t)Pw(t)\frac{dt}{t},$$

*with*

$$\int_0^\infty \widetilde{u}(t)\frac{dt}{t} = \int_0^\infty u(t)Pw(t)\frac{dt}{t};\ \int_0^\infty u(t)\frac{dt}{t} = 0,$$

*and*

$$\Phi_{\theta,q}(J(t, \widetilde{u}(t), \overline{X})) \leq c\,\|w\|_W\,\|f\|_{\bar{X}_{\theta,q}}$$

$$\Phi_{\theta,q}(J(t, u(t), \overline{X})) \leq c\,\|w\|_W\,\|f\|_{\bar{X}_{\theta,q}}$$

*Since*

$$\frac{1}{2}\int_0^\infty u(t)(Pw(t))^2\frac{dt}{t} = \frac{1}{2}\int_0^\infty (Pw(t))^2 d\left(\int_0^t u(s)\frac{ds}{s}\right)$$
$$= \int_0^\infty \left(\int_0^t u(s)\frac{ds}{s}\right) Pw(t)w^{\#}(t)\frac{dt}{t},$$

*it follows that if we let*

$$v(t) = (\int_0^t u(s)\frac{ds}{s})w^{\#}(t)$$

*then*

$$C_{2,Pw}(Tf) = \int_0^\infty (v(t) - \widetilde{u}(t))Pw(t)\frac{dt}{t},$$

*and*

$$\int_0^\infty (v(t) - \widetilde{u}(t))\frac{dt}{t} = 0.$$

*then theorem 2.1 implies that*

$$\|C_{2,Pw}(Tf)\|_{\bar{Y}_{\theta,q}} \leq c\,\|w\|_W\,\Phi_{\theta,q}(J(t,u(t);\bar{X})) + c\Phi_{\theta,q}(J(t,\widetilde{u}(t);\bar{X}))$$
$$\leq c\,\|w\|_W^2\,\|f\|_{\bar{X}_{\theta,q}}\,.$$

*as we wished to show.* □

## 5. Comparison with Earlier Results and Some Questions

This paper was originally conceived in 1999-2000, when the first named author spent one year in the Tropics. So publication was delayed somewhat and in the mean time several papers on the subject have appeared. In particular, [17] has similar statements framed in terms of weights of the form

$$w(t) = \phi(\log t), \text{ with } \phi \text{ Lipchitz.} \tag{15}$$

One recognizes that these weights are included in our theory since for $w$ of the form (15) we have (cf. Lemma 2.3 above)

$$\|w\|_{W_1} = \sup|tw'(t)| = \|\phi'\|_\infty < \infty.$$

There is also a connection with [1] (a longer version of this paper was originally circulated in 1996 (cf. [2])). These papers emphasize the connection between weighted norm inequalities, commutators and BMO type conditions using the $K-$method, and $BMO$ conditions are formulated in terms of properties of weights. Recall that for the $K-$method of interpolation we define the corresponding $\Omega$ operations by

$$\Omega^K f = \int_0^1 x_0(t)\frac{dt}{t} - \int_1^\infty x_1(t)\frac{dt}{t},$$

or, more generally, by

$$\Omega_w^K f = \int_0^1 x_0(t)w(t)\frac{dt}{t} - \int_1^\infty x_1(t)w(t)\frac{dt}{t},$$

where

$$f = x_0(t) + x_1(t), \text{ and } \|x_0(t)\|_{H_0} + t\,\|x_1(t)\|_{H_0} \leq cK(t, f; \bar{H}).$$

Using the strong form of the fundamental lemma of interpolation theory (cf. [6]) we can arrange to have $f = \int_0^\infty u_f(s)\frac{ds}{s}$, and

$$x_0(t) = \int_0^t u_f(s)\frac{ds}{s}, x_1(t) = \int_t^\infty u_f(s)\frac{ds}{s}.$$

It formally follows that

$$\Omega_w^K f = -\Omega_{Gw} f,$$

where

$$Gw(s) = \int_1^s w(r)\frac{dr}{r}.$$

In particular, if $w = 1$, then $Gw(s) = \log s$. Also note that

$$\sup_s |s(Gw)'(s)| = \|w\|_\infty .$$

Now a brief attempt to informally connect our work with Dan Waterman's classical Fourier analysis. One source of inspiration for the formulation of some of the results in this paper comes from the Littlewood-Paley theory, framed in terms of semigroups, e.g. as developed in Stein [21]. In the abstract theory of Stein [21] (cf. [21] pag 121) the relevant semigroups are represented, using the spectral theorem, by

$$T^t = \int_0^\infty e^{-\lambda t} dE(\lambda),$$

and one considers (multiplier) operators of the form

$$T_w f = \int_0^\infty e^{-\lambda t} w(t) dE(\lambda) f,$$

with $w \in L^\infty$. The conclusion is that the operator $T_{(Lw)'}$ is bounded on $L^p, 1 < p < \infty$, where

$$Lw(\lambda) = \int_0^\infty e^{-\lambda t} w(t) dt.$$

We hope to come back to explore this subject elsewhere.

We conclude with a few suggestions for future explorations on related topics.

T1. One can formulate iterations of the operation # (cf. [9]) and ask for its relevance in the study of higher order commutators.

T2. Despite several results (cf. [7], [2], [18]) one feels that the duality theory associated to the $\Omega$ operators is still not well developed. In particular, in [2] a predual $H$ of the space $W$ is constructed but the consequences have not been explored.

T3. Incidentally we note that the duality theory for the interpolation spaces introduced in [8] has not been studied.

T4. Compactness is a natural issue that has not been considered so far in abstract theory of commutators. For example, it is an important known result that commutators of CZO and functions in $VMO$ generate compact operators on $L^p$ (cf. [22]). We believe that the framework proposed in this paper could be useful to formulate the corresponding abstract result. In particular, one can define an appropriate analog of $VMO$...

T5. In connection with T3 and T4 it would be of interest to study compactness (weak compactness) in the abstract setting of [14] using the ideas in this paper.

## References

[1] J. Bastero, M. Milman and F. J. Ruiz, On the connection between weighted norm inequalities, commutators and real interpolation, Mem. Amer. Math. Soc. 731 (2001).

[2] J. Bastero, M. Milman and F. J. Ruiz, On the connection between weighted norm inequalities, commutators and real interpolation, Sem Galdeano (1996), Univ. Zaragoza, 101 pp.

[3] J. Bergh and J. Löfström, Interpolation Spaces. An Introduction, Springer-Verlag, Berlin/New York, 1976.

[4] C. Bennett, R. DeVore and R. Sharpley, Weak $L^\infty$ and $BMO$, Ann. of Math. 113 (1981), 601-611.

[5] R. Coifman, R. Rochberg and G. Weiss, Factorization theorems for Hardy spaces in several variables, Ann. of Math. 103 (1976), 611-635.

[6] M. Cwikel, B. Jawerth and M. Milman, On the fundamental lemma of interpolation theory, J. Approx. Th. 60 (1990), 70-82.

[7] M. Cwikel, B. Jawerth, M. Milman and R. Rochberg, Differential estimates and commutators in interpolation theory, in Analysis at Urbana, Volume II, pp 170-220, Cambridge University Press, 1989.

[8] M. Cwikel, N. Kalton, M. Milman and R. Rochberg, A unified theory of commutator estimates for a class of interpolation methods, Adv. Math. 169 (2002), 241-312.

[9] M. Cwikel and Y. Sagher, Weak type classes, J. Funct. Anal. 52 (1983), 11–18.

[10] C. Dellacherie, P. A. Meyer and M. Yor, Sur certaines propriétés des espaces de Banach $H^1$ et $BMO$, Sem. Prob. Strasbourg 12 (1978), 98-113.

[11] B. Jawerth, R. Rochberg and G. Weiss, Commutator and other second order estimates in real interpolation theory, Ark. Mat. 24 (1986), 191-219.

[12] N. Kalton, Non linear commutators in interpolation theory, Mem. Amer. Math. Soc 385 (1988).

[13] N. Kalton and S. Montgomery-Smith, Interpolation of Banach spaces, Handbook of the geometry of Banach spaces Vol. 2 (edited by W. B. Johnson and J. Lindenstrauss), Elsevier, 2003, pp 1163-1175.

[14] J. Martin and M. Milman, Modes of Convergence: Interpolation Methods I, J. Approx. Theory 111 (2001), 91-127.

[15] M. Milman and R. Rochberg, The role of cancellation in interpolation theory, Contemp. Math. 189 (1995), 403-419.

[16] M. Milman and Y. Sagher, An interpolation theorem, Ark. Mat. 22 (1984), 33-38.
[17] Ming Fan, Commutators in real interpolation with quasi-power parameters, Abstr. Appl. Anal. 7 (2002), 239-257.
[18] R. Rochberg, Uses of commutator theorems in analysis, Contemp. Math. 445 (2007), 277-295.
[19] R. Rochberg and G. Weiss, Derivatives of analytic families of Banach spaces, Ann. of Math. 118 (1983), 315-347.
[20] S. Semmes, Ode to commutators, Arxiv (http://arxiv.org/abs/math/0702462), 2007.
[21] E. M. Stein, Topics in harmonic analysis related to the LIttlewood-Paley theory, Annals of Math. Stud. 63, 1970, Princeton Univ. Press, Princeton.
[22] A. Uchiyama, Compactness of operators of Hankel type, Tôhoku Math. J. 30 (1978), 163-171.

# CONVERGENCE OF GREEDY APPROXIMATION WITH REGARD TO THE TRIGONOMETRIC SYSTEM *

V. TEMLYAKOV

*Department of Mathematics,*
*University of South Carolina,*
*Columbia, SC 29208,*
*USA*
*temlyak@math.sc.edu*

There has recently been much interest in approximation of functions by $m$-term approximants with regard to a basis (or minimal system). We discuss the following nonlinear method of approximation by trigonometric polynomials in this paper. For a periodic function $f$ we take as an approximant a trigonometric polynomial of the form $G_m(f) := \sum_{k\in\Lambda} \hat{f}(k)e^{i(k,x)}$, where $\Lambda \subset \mathbb{Z}^d$ is a set of cardinality $m$ containing the indices of the $m$ biggest (in absolute value) Fourier coefficients $\hat{f}(k)$ of function $f$. Note that $G_m(f)$ gives the best $m$-term approximant in the $L_2$-norm and, therefore, for each $f \in L_2$, $\|f - G_m(f)\|_2 \to 0$ as $m \to \infty$. It is known from previous results that in the case of $p \neq 2$ the condition $f \in L_p$ does not guarantee the convergence $\|f - G_m(f)\|_p \to 0$ as $m \to \infty$. The main goal of this paper is to complement a survey [11] by a discussion of recent results in the following setting: find an additional (to $f \in L_p$) condition on $f$ to guarantee that $\|f - G_m(f)\|_p \to 0$ as $m \to \infty$.

## 1. Introduction

Let a Banach space $X$ with a system (for instance, a basis) $\Psi = \{\psi_k\}_{k=1}^{\infty}$, be given. We assume that for each $f \in X$ there is a unique series

$$f \sim \sum_{k=1}^{\infty} c_k(f, \Psi)\psi_k \tag{1.1}$$

with the property

$$\lim_{k\to\infty} c_k(f, \Psi) = 0.$$

We study the following theoretical greedy algorithm. For a given element $f \in X$ we consider the expansion (1.1). For an element $f \in X$ we call a permutation $\rho$, $\rho(j) = k_j$, $j = 1, 2, \dots$, of the positive integers decreasing and write $\rho \in D(f)$ if

$$|c_{k_1}(f, \Psi)| \geq |c_{k_2}(f, \Psi)| \geq \dots \quad . \tag{1.2}$$

*This research has been supported by NSF Grant DMS 0554832.

In the case of strict inequalities here $D(f)$ consists of only one permutation. We define the $m$-th greedy approximant of $f$ with regard to the system $\Psi$ corresponding to a permutation $\rho \in D(f)$ by formula

$$G_m(f) := G_m(f, \Psi) := G_m(f, \Psi, \rho) := \sum_{j=1}^{m} c_{k_j}(f, \Psi)\psi_{k_j}.$$

The greedy approximant $G_m(f, \Psi)$ is a partial sum of the rearranged series

$$\sum_{j=1}^{\infty} c_{k_j}(f, \Psi)\psi_{k_j}. \tag{1.3}$$

An immediate question with (1.3) is when does this series converge? The theory of convergence of rearranged series is a classical topic in analysis. A series converges *unconditionally* if every rearrangement of this series converges. A system (basis) $\Psi$ of a Banach space $X$ is said to be an *unconditional system (basis)* if for every $f \in X$ its expansion (1.1) converges unconditionally. For a set of indices $\Lambda$ define

$$S_\Lambda(f, \Psi) := \sum_{n \in \Lambda} c_n(f, \Psi)\psi_n.$$

It is well known that if $\Psi$ is unconditional then there exists a constant $K$ such that for any $\Lambda$

$$\|S_\Lambda(f, \Psi)\| \le K\|f\|. \tag{1.4}$$

This inequality implies

$$\|f - S_\Lambda(f, \Psi)\| \le (K+1)E_\Lambda(f, \Psi) \tag{1.5}$$

where

$$E_\Lambda(f, \Psi) := \inf_{\{c_n\}} \|f - \sum_{n \in \Lambda} c_n\psi_n\|.$$

The inequality (1.5) indicates that in the case of an unconditional basis $\Psi$ it is sufficient for finding near best $m$-term approximant to optimize only over the sets of indices $\Lambda$. The greedy algorithm $G_m(\cdot, \Psi)$ gives a simple recipe for building $\Lambda_m$: pick the indices with biggest coefficients. In [6] we discuss in detail when the above simple recipe provides a near best $m$-term approximant. It turned out that the only assumption that $\Psi$ is unconditional does not guarantee that $G_m(\cdot, \Psi)$ provides a near best $m$-term approximation. In [6] (for further discussion see a survey [8]) we discuss a new class of bases (*greedy bases*) that has the property that $G_m(f, \Psi)$ provides a near best $m$-term approximation for each $f \in X$. We showed in [6] that the class of greedy bases is a proper subclass of the class of unconditional bases.

It follows from the definition of unconditional basis that any rearrangement of the series in (1.1) converges. It is known that it converges to $f$. The rearrangement (1.3) is a specific rearrangement of (1.1). Clearly, for an unconditional basis $\Psi$ (1.3) converges to $f$. It turned out that unconditionality of $\Psi$ is not a necessary condition for convergence of (1.3) for each $f \in X$. Bases that have the property of convergence of (1.3) for each $f \in X$ are exactly the *quasi-greedy bases* (see [6], [8]).

The first results on greedy approximation with regard to bases showed that the Haar basis and other bases similar to it are very well designed for greedy approximation (see, for instance, survey [8]). In this paper (see Section 2) we discuss another classical system, namely, the trigonometric system from the point of view of greedy approximation. It is well known that the trigonometric system is not an unconditional basis for $L_p$, $p \neq 2$. Therefore, by [6] it is not a greedy basis for $L_p$, $p \neq 2$. In this paper we mostly discuss convergence properties of the Weak Greedy Algorithm with regard to the trigonometric system. It is a nontrivial problem. In Section 2 we will demonstrate how it relates to some deep results in harmonic and functional analysis.

## 2. Greedy Approximation with Regard to the Trigonometric System

Consider a periodic function $f \in L_p(\mathbb{T}^d)$, $1 \leq p \leq \infty$, $(L_\infty(\mathbb{T}^d) = \mathcal{C}(\mathbb{T}^d))$, defined on the $d$-dimensional torus $\mathbb{T}^d$. Let a number $m \in \mathbb{N}$ and a number $t \in (0,1]$ be given and $\Lambda_m$ be a set of $k \in \mathbb{Z}^d$ with the properties:

$$\min_{k\in\Lambda_m} |\hat{f}(k)| \geq t \max_{k\notin\Lambda_m} |\hat{f}(k)|, \quad |\Lambda_m| = m, \tag{2.1}$$

where

$$\hat{f}(k) := (2\pi)^{-d} \int_{\mathbb{T}^d} f(x)e^{-i(k,x)}dx$$

is a Fourier coefficient of $f$. We define

$$G_m^t(f) := G_m^t(f, \mathcal{T}^d) := S_{\Lambda_m}(f) := \sum_{k\in\Lambda_m} \hat{f}(k)e^{i(k,x)}$$

and call it an $m$-th weak greedy approximant of $f$ with regard to the trigonometric system $\mathcal{T}^d := \{e^{i(k,x)}\}_{k\in\mathbb{Z}^d}$, $\mathcal{T} := \mathcal{T}^1$. We write $G_m(f) = G_m^1(f)$ and call it an $m$-th greedy approximant. Clearly, an $m$-th weak greedy approximant and even an $m$-th greedy approximant may not be unique. In this section we do not impose any extra restrictions on $\Lambda_m$ in addition to (2.1). Thus theorems formulated below hold for any choice of $\Lambda_m$ satisfying (2.1) or in other words for any realization $G_m^t(f)$ of the weak greedy approximation.

We will discuss in detail only results concerning convergence of the weak greedy approximants with regard to the trigonometric system. T.W. Körner answering a question raised by Carleson and Coifman constructed in [4] a function from $L_2(\mathbb{T})$ and then in [5] a continuous function such that $\{G_m(f,\mathcal{T})\}$ diverges almost everywhere. It has been proved in [10] for $p \neq 2$ and in [2] for $p < 2$ that there exists a $f \in L_p(\mathbb{T})$ such that $\{G_m(f,\mathcal{T})\}$ does not converge in $L_p$. It was remarked in [11] that the method from [10] gives a little more: 1) There exists a continuous function $f$ such that $\{G_m(f,\mathcal{T})\}$ does not converge in $L_p(\mathbb{T})$ for any $p > 2$; 2) There exists a function $f$ that belongs to any $L_p(\mathbb{T})$, $p < 2$, such that $\{G_m(f, \mathcal{T})\}$ does not converge in measure. Thus the above negative results show that the condition $f \in L_p(\mathbb{T}^d)$, $p \neq 2$, does not guarantee convergence of $\{G_m(f, \mathcal{T})\}$ in the $L_p$-norm. The main goal of this paper is to complement a survey [11] by recent results in the following setting: find an additional (to $f \in L_p$) condition on $f$ to guarantee that $\|f - G_m(f, \mathcal{T})\|_p \to 0$ as $m \to \infty$. In [7] we proved the following theorem.

**Theorem 2.1.** *Let $f \in L_p(\mathbb{T}^d)$, $2 < p \leq \infty$, and let $q > p' := p/(p-1)$. Assume that $f$ satisfies the condition*

$$\sum_{|k|>n} |\hat{f}(k)|^q = o(n^{d(1-q/p')})$$

*where $|k| := \max_{1 \leq j \leq d} |k_j|$. Then we have*

$$\lim_{m\to\infty} \|f - G_m^t(f, \mathcal{T})\|_p = 0.$$

It is proved in [7] that Theorem 2.1 is sharp.

**Proposition 2.1.** *For each $2 < p \leq \infty$ there exists $f \in L_p(\mathbb{T}^d)$ such that*

$$|\hat{f}(k)| = O(|k|^{-d(1-1/p)}),$$

*and the sequence $\{G_m(f)\}$ diverges in the $L_p$.*

Let us make some comments. For a given set $\Lambda$ denote

$$E_\Lambda(f)_p := \inf_{c_k, k\in\Lambda} \|f - \sum_{k\in\Lambda} c_k e^{i(k,x)}\|_p, \quad S_\Lambda(f) := \sum_{k\in\Lambda} \hat{f}(k) e^{i(k,x)}.$$

Define a special domain

$$Q(n) := \{k : |k| \leq n^{1/d}\}.$$

**Remark 2.1.** Theorem 2.1 implies that if $f \in L_p$, $2 < p \leq \infty$, and

$$E_{Q(n)}(f)_2 = o(n^{1/p-1/2})$$

then $G_m^t(f) \to f$ in $L_p$.

**Remark 2.2.** The proof of Proposition 2.1 (see [7]) implies that there is a $f \in L_p(\mathbb{T}^d)$ such that

$$E_{Q(n)}(f)_\infty = O(n^{1/p-1/2})$$

and $\{G_m(f)\}$ diverges in $L_p$, $2 < p \leq \infty$.

We note that Remark 2.1 can also be obtained from some general inequalities for $\|f - G_m^t(f)\|_p$. We define the best $m$-term approximation with regard to $\mathcal{T}^d$

$$\sigma_m(f)_p := \sigma_m(f, \mathcal{T}^d)_p := \inf_{k^j \in \mathbb{Z}^d, c_j} \|f - \sum_{j=1}^{m} c_j e^{i(k^j,x)}\|_p.$$

The following inequality has been proved in [10] for $t = 1$ and in [7] for general $t$.

**Theorem 2.2.** *For each $f \in L_p(\mathbb{T}^d)$ and any $0 < t \le 1$ we have*

$$\|f - G_m^t(f)\|_p \le (1 + (2 + 1/t)m^{h(p)})\sigma_m(f)_p, \quad 1 \le p \le \infty, \tag{2.2}$$

*where $h(p) := |1/2 - 1/p|$.*

It was proved in [10] that the inequality (2.2) is sharp: there is a positive absolute constant $C$ such that for each $m$ and $1 \le p \le \infty$ there exists a function $f \neq 0$ with the property

$$\|G_m(f)\|_p \ge Cm^{h(p)}\|f\|_p. \tag{2.3}$$

The above inequality (2.3) shows that the trigonometric system is not a quasi-greedy basis for $L_p$, $p \neq 2$ (see [6] for the definition of a quasi-greedy basis). We formulate one more inequality from [7].

**Theorem 2.3.** *Let $2 \le p \le \infty$. Then for any $f \in L_p(\mathbb{T}^d)$ and any $Q$, $|Q| \le m$, we have*

$$\|f - G_m^t(f)\|_p \le \|f - S_Q(f)\|_p + (3 + 1/t)(2m)^{h(p)}E_Q(f)_2.$$

We present some results from [7] that are formulated in terms of the Fourier coefficients. For $f \in L_1(\mathbb{T}^d)$ let $\{\hat{f}(k(l))\}_{l=1}^{\infty}$ denote the decreasing rearrangement of $\{\hat{f}(k)\}_{k \in \mathbb{Z}^d}$, i.e.

$$|\hat{f}(k(1))| \ge |\hat{f}(k(2))| \ge \dots \quad .$$

Denote $a_n(f) := |\hat{f}(k(n))|$.

**Theorem 2.4.** *Let $2 < p < \infty$ and let a decreasing sequence $\{A_n\}_{n=1}^{\infty}$ satisfy the condition:*

$$A_n = o(n^{1/p-1}) \quad as \quad n \to \infty.$$

*Then for any $f \in L_p(\mathbb{T}^d)$ with the property $a_n(f) \le A_n$, $n = 1, 2, \dots$, we have*

$$\lim_{m\to\infty} \|f - G_m^t(f)\|_p = 0.$$

We also proved in [7] that for any decreasing sequence $\{A_n\}$, satisfying

$$\limsup_{n\to\infty} A_n n^{1-1/p} > 0$$

there exists a function $f \in L_p$ such that $a_n(f) \le A_n$, $n = 1, \dots$, with divergent in the $L_p$ sequence of greedy approximants $\{G_m(f)\}$.

In [7] we proved a necessary and sufficient condition on the majorant $\{A_n\}$ to guarantee (under assumption that $f$ is a continuous function) uniform convergence of greedy approximants to a function $f$.

**Theorem 2.5.** *Let a decreasing sequence $\{A_n\}_{n=1}^{\infty}$ satisfy the condition $(\mathcal{A}_\infty)$:*

$$\sum_{M<n\le e^M} A_n = o(1) \quad as \quad M \to \infty.$$

*Then for any $f \in \mathcal{C}(\mathbb{T})$ with the property $a_n(f) \le A_n$, $n = 1, 2, \dots,$ we have*

$$\lim_{m\to\infty} \|f - G_m^t(f, \mathcal{T})\|_\infty = 0.$$

The condition $(\mathcal{A}_\infty)$ is very close to the convergence of the series $\sum_n A_n$; if the condition $(\mathcal{A}_\infty)$ holds then we have

$$\sum_{n=1}^{N} A_n = o(\log_*(N)), \quad \text{as} \quad N \to \infty,$$

where a function $\log_*(u)$ is defined to be bounded for $u \le 0$ and to satisfy $\log_*(u) = \log_*(\log u) + 1$ for $u > 0$. The function $\log_*(u)$ grows slower than any iterated logarithmic function.

The condition $(\mathcal{A}_\infty)$ in Theorem 2.5 is sharp.

**Theorem 2.6.** *Assume that a decreasing sequence $\{A_n\}_{n=1}^{\infty}$ does not satisfy the condition $(\mathcal{A}_\infty)$. Then there exists a function $f \in \mathcal{C}(\mathbb{T})$ with the property $a_n(f) \le A_n$, $n = 1, 2, \dots,$ and such that we have*

$$\limsup_{m\to\infty} \|f - G_m(f, \mathcal{T})\|_\infty > 0$$

*for some realization $G_m(f, \mathcal{T})$.*

In [9] we concentrated on imposing extra conditions in the following form. We assume that for some sequence $\{M(m)\}$, $M(m) > m$, we have

$$\|G_{M(m)}(f) - G_m(f)\|_p \to 0 \quad \text{as} \quad m \to \infty.$$

In the case $p$ is an even number or $p = \infty$ we found in [9] necessary and sufficient conditions on the growth of the sequence $\{M(m)\}$ to provide convergence $\|f - G_m(f)\|_p \to 0$ as $m \to \infty$. We proved the following theorem in [9].

**Theorem 2.7.** *Let $p = 2q$, $q \in \mathbb{N}$, be an even integer, $\delta > 0$. Assume that $f \in L_p(\mathbb{T})$ and there exists a sequence of positive integers $M(m) > m^{1+\delta}$ such that*

$$\|G_m(f) - G_{M(m)}(f)\|_p \to 0 \quad as \quad m \to \infty.$$

*Then we have*

$$\|G_m(f) - f\|_p \to 0 \quad as \quad m \to \infty.$$

In [9] we proved that the condition $M(m) > m^{1+\delta}$ cannot be replaced by a condition $M(m) > m^{1+o(1)}$.

**Theorem 2.8.** *For any $p \in (2, \infty)$ there exists a function $f \in L_p(\mathbb{T})$ with divergent in the $L_p(\mathbb{T})$ sequence $\{G_m(f)\}$ of greedy approximations with the following property. For any sequence $\{M(m)\}$ such that $m \le M(m) \le m^{1+o(1)}$ we have*

$$\|G_{M(m)}(f) - G_m(f)\|_p \to 0 \quad (m \to 0).$$

In [9] we also considered the case $p = \infty$. We proved there necessary and sufficient conditions for convergence of greedy approximations in the uniform norm. For a mapping $\alpha : W \to W$ we denote $\alpha_k$ its $k$-fold iteration: $\alpha_k := \alpha \circ \alpha_{k-1}$.

**Theorem 2.9.** *Let $\alpha : \mathbb{N} \to \mathbb{N}$ be strictly increasing. Then the following conditions are equivalent:*
*a) for some $k \in \mathbb{N}$ and for any sufficiently large $m \in \mathbb{N}$ we have $\alpha_k(m) > e^m$;*
*b) if $f \in C(\mathbb{T})$ and*

$$\left\|G_{\alpha(m)}(f) - G_m(f)\right\|_\infty \to 0 \quad (m \to \infty)$$

*then*

$$\|f - G_m(f)\|_\infty \to 0 \quad (m \to \infty).$$

In order to illustrate some technique used in proofs of the above results we discuss some inequalities that were used in proving Theorems 2.7 and 2.9. The reader will also see from the further discussion a connection to some deep results in harmonic analysis. The general style of these inequalities is the following. A function that has a sparse representation with regard to the trigonometric system cannot be approximated in $L_p$ by functions with small Fourier coefficients. We begin our discussion with some concepts introduced in [9] that are useful in proving such inequalities. The following new characteristic of a Banach space $L_p$ plays an important role in such inequalities. We introduce some more notations. Let $\Lambda$ be a finite subset of $\mathbb{Z}^d$. By $|\Lambda|$ we denote its cardinality and by $\mathcal{T}(\Lambda)$ the span of $\{e^{i(k,x)}\}_{k\in\Lambda}$. Denote

$$\Sigma_m(\mathcal{T}) = \cup_{\Lambda:|\Lambda|\le m}\mathcal{T}(\Lambda).$$

For $f \in L_p$, $F \in L_{p'}$, $1 \le p \le \infty$, $p' = p/(p-1)$, we write

$$\langle F, f\rangle := \int_{\mathbb{T}^d} F\bar{f}d\mu, \quad d\mu := (2\pi)^{-d}dx.$$

**Definition 2.1.** Let $\Lambda$ be a finite subset of $\mathbb{Z}^d$ and $1 \le p \le \infty$. We call a set $\Lambda' := \Lambda'(p,\gamma)$, $\gamma \in (0,1]$ a $(p,\gamma)$-dual to $\Lambda$ if for any $f \in \mathcal{T}(\Lambda)$ there exists $F \in \mathcal{T}(\Lambda')$ such that $\|F\|_{p'} = 1$ and $\langle F, f\rangle \ge \gamma\|f\|_p$.

Denote by $D(\Lambda, p, \gamma)$ the set of all $(p,\gamma)$-dual sets $\Lambda'$. The following function is important for us

$$v(m,p,\gamma) := \sup_{\Lambda:|\Lambda|=m} \inf_{\Lambda'\in D(\Lambda,p,\gamma)} |\Lambda'|.$$

We note that in a particular case $p = 2q$, $q \in \mathbb{N}$ we have

$$v(m,p,1) \le m^{p-1}. \tag{2.4}$$

This follows immediately from the form of the norming functional $F$ for $f \in L_p$:

$$F = f^{q-1}(\bar{f})^q\|f\|_p^{1-p}. \tag{2.5}$$

In [9] we used the quantity $v(m,p,\gamma)$ in greedy approximation. We first prove a lemma.

**Lemma 2.1.** *Let $2 \le p \le \infty$. For any $h \in \Sigma_m(\mathcal{T})$ and any $g \in L_p$ one has*

$$\|h+g\|_p \ge \gamma\|h\|_p - v(m,p,\gamma)^{1-1/p}\|\{\hat{g}(k)\}\|_{\ell_\infty}.$$

**Proof.** Let $h \in \mathcal{T}(\Lambda)$ with $|\Lambda| = m$ and let $\Lambda' \in D(\Lambda,p,\gamma)$. Then using the Definition 2.1 we find $F(h,\gamma) \in \mathcal{T}(\Lambda')$ such that

$$\|F(h,\gamma)\|_{p'} = 1 \quad \text{and} \quad \langle F(h,\gamma),h\rangle \ge \gamma\|h\|_p.$$

We have

$$\langle F(h,\gamma),h\rangle = \langle F(h,\gamma),h+g\rangle - \langle F(h,\gamma),g\rangle \le \|h+g\|_p + |\langle F(h,\gamma),g\rangle|.$$

Next,

$$|\langle F(h,\gamma),g\rangle| \le \|\{\hat{F}(h,\gamma)(k)\}\|_{\ell_1}\|\{\hat{g}(k)\}\|_{\ell_\infty}.$$

Using $F(h,\gamma) \in \mathcal{T}(\Lambda')$ and the Hausdorf-Young theorem [12], Ch.12, Section 2, we obtain

$$\|\{\hat{F}(h,\gamma)(k)\}\|_{\ell_1} \le |\Lambda'|^{1-1/p}\|\{\hat{F}(h,\gamma)(k)\}\|_{\ell_p}$$

$$\le |\Lambda'|^{1-1/p}\|F(h,\gamma)\|_{p'} = |\Lambda'|^{1-1/p}.$$

It remains to combine the above inequalities and use the definition of $v(m,p,\gamma)$. □

**Definition 2.2.** Let $X$ be a finite dimensional subspace of $L_p$, $1 \le p \le \infty$. We call a subspace $Y \subset L_{p'}$ a $(p,\gamma)$-dual to $X$, $\gamma \in (0,1]$, if for any $f \in X$ there exists $F \in Y$ such that $\|F\|_{p'} = 1$ and $\langle F,f\rangle \ge \gamma\|f\|_p$.

Similarly to the above denote by $D(X,p,\gamma)$ the set of all $(p,\gamma)$-dual subspaces $Y$. Consider the following function

$$w(m,p,\gamma) := \sup_{X:\dim X = m} \inf_{Y \in D(X,p,\gamma)} \dim Y.$$

We begin our discussion by a particular case $p = 2q$, $q \in \mathbb{N}$. Let $X$ be given and $e_1,\dots,e_m$ form a basis of $X$. Using the Hölder inequality for $n$ functions $f_1,\dots,f_n \in L_n$

$$\int |f_1 \cdots f_n| d\mu \le \|f_1\|_n \cdots \|f_n\|_n$$

with $f_i = |e_j|^{p'}$, $n = p-1$ we get that any function of the form

$$\prod_{i=1}^m |e_i|^{k_i}, \quad k_i \in \mathbb{N}, \quad \sum_{i=1}^m k_i = p-1,$$

belongs to $L_{p'}$. It now follows from (2.5) that

$$w(m,p,1) \le m^{p-1}, \quad p = 2q, \quad q \in \mathbb{N}. \tag{2.6}$$

There is a general theory of uniform approximation property (UAP) that provides some estimates for $w(m,p,\gamma)$ and $v(m,p,\gamma)$. We give some definitions from this theory. For a given subspace $X$ of $L_p$, $\dim X = m$, and a constant $K > 1$ let $k_p(X,K)$ be the smallest $k$ such that there is an operator $I_X : L_p \to L_p$ with $I_X(f) = f$ for $f \in X$, $\|I_X\|_{L_p\to L_p} \le K$, and $\mathrm{rank} I_X \le k$. Denote

$$k_p(m,K) := \sup_{X:\dim X = m} k_p(X,K).$$

Let us discuss how $k_p(m,K)$ can be used in estimating $w(m,p,\gamma)$. Consider $I_X^*$ the dual to $I_X$ operator. Then $\|I_X^*\|_{L_{p'}\to L_{p'}} \le K$ and $\mathrm{rank} I_X^* \le k_p(m,K)$. Let $f \in X$, $\dim X = m$, and let $F_f$ be the norming functional for $f$. Define

$$F := I_X^*(F_f)/\|I_X^*(F_f)\|_{p'}.$$

Then $(f \in X)$

$$\langle f, I_X^*(F_f)\rangle = \langle I_X(f), F_f\rangle = \langle f, F_f\rangle = \|f\|_p$$

and

$$\|I_X^*(F_f)\|_{p'} \le K$$

imply

$$\langle f, F\rangle \ge K^{-1}\|f\|_p.$$

Therefore

$$w(m,p,K^{-1}) \le k_p(m,K). \tag{2.7}$$

We note that the behavior of functions $w(m,p,\gamma)$ and $k_p(m,K)$ may be very different. J. Bourgain [1] proved that for any $p \in (1,\infty)$, $p \ne 2$ the function $k_p(m,K)$ grows faster than any polynomial in $m$. The estimate (2.6) shows that in the particular case $p = 2q$, $q \in \mathbb{N}$ the growth of $w(m,p,\gamma)$ is at most polynomial. This means that we cannot expect to obtain accurate estimates for $w(m,p,K^{-1})$ using the inequality (2.7). We give one more application of the UAP in the style of Lemma 2.1.

**Lemma 2.2.** *Let $2 \le p \le \infty$. For any $h \in \Sigma_m(\mathcal{T})$ and any $g \in L_p$ one has*

$$\|h+g\|_p \ge K^{-1}\|h\|_p - k_p(m,K)^{1/2}\|g\|_2; \tag{2.8}$$

$$\|h+g\|_p \ge K^{-2}\|h\|_p - k_p(m,K)\|\{\hat{g}(k)\}\|_{\ell_\infty}. \tag{2.9}$$

**Proof.** Let $h \in \mathcal{T}(\Lambda)$, $|\Lambda| = m$. Take $X = \mathcal{T}(\Lambda)$ and consider the operator $I_X$ provided by the UAP. Let $\psi_1,\dots,\psi_M$ form an orthonormal basis for the range $Y$ of the operator $I_X$. Then $M \le k_p(m,K)$. Let

$$I_X(e^{i(k,x)}) = \sum_{j=1}^{M} c_j^k \psi_j.$$

Then the property $\|I_X\|_{L_p\to L_p} \le K$ implies

$$(\sum_{j=1}^{M} |c_j^k|^2)^{1/2} = \|I_X(e^{i(k,x)})\|_2 \le \|I_X(e^{i(k,x)})\|_p \le K.$$

Consider along with the operator $I_X$ a new one

$$A := (2\pi)^{-d}\int_{T^d} T_t I_X T_{-t} dt$$

where $T_t$ is a shifting operator: $T_t(f) = f(\cdot + t)$. Then

$$A(e^{i(k,x)}) = \sum_{j=1}^{M} c_j^k (2\pi)^{-d}\int_{T^d} e^{-i(k,t)}\psi_j(x+t)dt = (\sum_{j=1}^{M} c_j^k \hat{\psi}_j(k))e^{i(k,x)}.$$

Denote

$$\lambda_k := \sum_{j=1}^{M} c_j^k \hat{\psi}_j(k).$$

We have

$$\sum_k |\lambda_k|^2 \le \sum_k (\sum_{j=1}^{M} |c_j^k|^2)(\sum_{j=1}^{M} |\hat{\psi}(k)|^2) \le K^2 M.$$

Also $\lambda_k = 1$ for $k \in \Lambda$. For the operator $A$ we have

$$\|A\|_{L_p\to L_p} \le K \quad \text{and} \quad \|A\|_{L_2\to L_\infty} \le KM^{1/2}.$$

Therefore

$$\|A(h+g)\|_p \le K\|h+g\|_p$$

and

$$\|A(h+g)\|_p \ge \|h\|_p - KM^{1/2}\|g\|_2.$$

This proves the first inequality.

Consider the operator $B := A^2$. Then

$$B(h) = h, \quad h \in \mathcal{T}(\Lambda); \quad \|B\|_{L_p\to L_p} \le K^2$$

and

$$\|B(f)\|_\infty \le K^2 M\|\{\hat{f}(k)\}\|_{\ell_\infty}.$$

Now, on the one hand

$$\|B(h+g)\|_p \le K^2\|h+g\|_p$$

and on the other hand

$$\|B(h+g)\|_p = \|h + B(g)\|_p \ge \|h\|_p - K^2 M\|\{\hat{g}(k)\}\|_{\ell_\infty}.$$

This proves inequality (2.9). □

**Theorem 2.10.** *For any $h \in \Sigma_m(\mathcal{T})$ and any $g \in L_\infty$ one has*

$$\|h+g\|_\infty \geq K^{-1}\|h\|_\infty - e^{C(K)m/2}\|g\|_2;$$

$$\|h+g\|_\infty \geq K^{-2}\|h\|_\infty - e^{C(K)m}\|\{\hat{g}(k)\}\|_{\ell_\infty}.$$

**Proof.** This theorem is a direct corollary of Lemma 2.2 and the following known (see [3]) estimate

$$k_\infty(m,K) \leq e^{C(K)m}. \qquad \square$$

As we already mentioned $k_p(m,K)$ increases faster than any polynomial. In [9] we improved inequality (2.8) by using other arguments.

**Lemma 2.3.** *Let $2 \leq p \leq \infty$. For any $h \in \Sigma_m(\mathcal{T})$ and any $g \in L_p$ one has*

$$\|h+g\|_p^p \geq 2^{-p-1}\|h\|_p^p - 2m^{p/2}\|h\|_p^{p-2}\|g\|_2^2. \tag{2.10}$$

We mention two inequalities from [9] in a style of inequalities in Lemmas 2.1–2.3.

**Lemma 2.4.** *Let $2 \leq p < \infty$ and $h \in L_p$, $\|h\|_p \neq 0$. Then for any $g \in L_p$ we have*

$$\|h\|_p \leq \|h+g\|_p + (\|h\|_{2p-2}/\|h\|_p)^{p-1}\|g\|_2.$$

**Lemma 2.5.** *Let $h \in \Sigma_m(\mathcal{T})$, $\|h\|_\infty = 1$. Then for any function $g$ such that $\|g\|_2 \leq \frac{1}{4}(4\pi m)^{-m/2}$ we have*

$$\|h+g\|_\infty \geq 1/4.$$

In the special case of even $p$ we have by (2.4) and (2.6) that

$$v(m,p,1) \leq m^{p-1}, \quad w(m,p,1) \leq m^{p-1}.$$

The following bound has been proved in [9].

**Lemma 2.6.** *Let $2 \leq p < \infty$. Denote $\alpha := p/2 - [p/2]$. Then we have*

$$v(m,p,\gamma) \leq m^{c(\alpha,\gamma)m^{1/2}+p-1}.$$

## References


[1] J. Bourgain, A remark on the behaviour of $L^p$-multipliers and the range of operators acting on $L^p$-spaces, *Israel J. Math.*, **79**(1992), 193–206.

[2] A. Cordoba and P. Fernandez, Convergence and divergence of decreasing rearranged Fourier series, *SIAM, I. Math. Anal.*, **29**(1998), 1129–1139.

[3] T. Figiel, W.B. Johnson, G. Schechtman, Factorization of natural embeddings of $\ell_p^n$ into $L_r$, I, *Studia Mathematica*, **89**(1988), 79–103.

[4] T.W. Körner, Divergence of decreasing rearranged Fourier series, *Annals of Mathematics*, **144**(1996), 167–180.

[5] T.W. Körner, Decreasing rearranged Fourier series, *The J. Fourier Analysis and Applications*, **5**(1999), 1–19.

[6] S.V. Konyagin and V.N. Temlyakov, A remark on greedy approximation in Banach spaces, *East J. on Approx.*, **5**(1999), 1–15.

[7] S.V. Konyagin and V.N. Temlyakov, Convergence of Greedy Approximation II. The Trigonometric System, *Studia Mathematica*, **159(2)**(2003), 161–184.

[8] S.V. Konyagin and V.N. Temlyakov, Greedy Approximation with Regard to bases and General Minimal Systems, *Serdica Math. J.*, **28**(2002), 305–328.

[9] S.V. Konyagin and V.N. Temlyakov, Convergence of greedy approximation for the trigonometric system, *Analysis Mathematica*, **31**(2005), 85–115.

[10] V.N. Temlyakov, Greedy Algorithm and $m$-term Trigonometric Approximation, *Constructive Approximation*, **14**(1998), 569–582.

[11] V.N. Temlyakov, Nonlinear methods of approximation, *Found. Comput. Math.*, **3**(2003), 33–102.

[12] A. Zygmund, *Trigonometric series*, (University Press, Cambridge, 1959).

# FUNCTIONS OF BOUNDED Λ-VARIATION

FRANCISZEK PRUS-WIŚNIOVSKI

*Instytut Matematyki*
*Uniwersytet Szczeciński*
*ul. Wielkopolska 15*
*70-453 Szczecin*
*Poland*
*wisniows@univ.szczecin.pl*

This paper surveys the results on general properties of functions of bounded Λ-variation in one variable, including some not yet published. The extensions to several variables and applications to Fourier series are not discussed. The relation between the class $\Lambda BV$ of functions of bounded Λ-variation and its important subclass $\Lambda BV_c$ of functions continuous in lambda-variation is discussed in detail. Open problems are described.

*Keywords*: Generalized variation; generalized absolute continuity

## 1. About the Definition

The concept of bounded harmonic variation arose naturally from the theory of Fourier series and has many applications to it. The most important among them is the following theorem that furnishes the same conclusions as the Dirichlet-Jordan theorem [35, Thm. 2].

**Theorem 1.1 (Waterman's Test).** *If $f$ is of bounded harmonic variation, then*

*(i)* $S[f](x)$ *converges to* $\frac{f(x+)+f(x-)}{2}$ *pointwise;*
*(ii)* $S[f]$ *converges to* $f$ *uniformly on every closed interval of points of continuity.*

E. I. Berezhnoi showed that Waterman's Test is the strongest of all tests for uniform convergence of Fourier series based on functions of generalized bounded variation [3, Thm. 12] and that it cannot be improved [3, Thm.13].

A sequence $\Lambda = (\lambda_i)$ is said to be a Λ-sequence if it is a non-decreasing sequence of positive numbers such that $\sum \frac{1}{\lambda_i}$ diverges. Two most important examples of Λ-sequences are given by $J = (1)_{i=1}^{\infty}$ and $H := (i)_{i=1}^{\infty}$. A Λ-sequence $\Lambda = (\lambda_i)$ is said to be proper if $\lim \lambda_i = +\infty$. Thus, the Λ-sequence $H$ is proper, while $J$ is not.

Given a $\Lambda$-sequence $\Lambda = (\lambda_i)$, a function $f : [a, b] \to \mathbb{R}$ is called to be of bounded $\Lambda$-variation if

$$\sum_{i=1}^{\infty} \frac{|f(I_i)|}{\lambda_i} < +\infty$$

for every collection $\mathcal{I} = \{I_i\}_{i=1}^{\infty}$ of non-overlapping closed subintervals $I_i \subset [a, b]$, where $f(I_i)$ is the signed change of $f$ over the interval $I_i$, that is, $f(I_i) := f(\sup I_i) - f(\inf I_i)$ [35]. Since reordering of the collection $\{I_i\}_{i=1}^{\infty}$ usually results in a change of the value of $\sum_i \frac{|f(I_i)|}{\lambda_i}$, it will be convenient to introduce the following symbol

$$\sigma_\Lambda(f, \mathcal{I}) := \sup_{\beta} \sum_{I \in \mathcal{I}} \frac{|f(I)|}{\lambda_{\beta(I)}}$$

where the supremum is taken over all injective mappings $\beta : \mathcal{I} \to \mathbb{N}$. This yields a quantity that depends on the elements of the collection $\mathcal{I}$ only, and not on their enumeration. The following observation [37, Thm. 1] is fundamental for establishing a relation between the new concept of functions of bounded $\Lambda$-variation and the well-known functions of bounded (Jordan or ordinary) variation.

**Theorem 1.2.** *The following statements are equivalent:*

*(i) $f$ is of bounded $\Lambda$-variation;*
*(ii) there is an $M < +\infty$ such that*

$$\sum_{i=1}^{\infty} \frac{|f(I_i)|}{\lambda_i} < M$$

*for every sequence $\{I_i\}_{i=1}^{\infty}$ of non-overlapping closed subintervals of $[a, b]$;*
*(iii) there is an $M < +\infty$ such that*

$$\sum_{i=1}^{n} \frac{|f(I_i)|}{\lambda_i} < M$$

*for every finite collection $\{I_i\}_{i=1}^{n}$ of non-overlapping closed subintervals of $[a, b]$.*

The infimum of such $M$ is called the $\Lambda$-variation of $f$ and denoted by $V_\Lambda(f)$ (or by $V_\Lambda(f, [a, b])$ if the interval $[a, b]$ is not the entire domain of $f$). Thus, defining

$$V_\Lambda(f) := \sup_{\mathcal{I}} \sigma_\Lambda(f, \mathcal{I}),$$

where the supremum is taken over all possible families $\mathcal{I}$ of non-overlapping closed subintervals of $[a, b]$, we can say that a function $f$ is of bounded $\Lambda$-variation if and only if $V_\Lambda(f)$ is finite. The quantity $V_\Lambda(f)$ is called the $\Lambda$-variation of $f$. The best way of thinking about $\sigma_\Lambda(f, \mathcal{I})$ is to perceive the family $\mathcal{I}$ as $f$-ordered (that is, $\mathcal{I} = \{I_1, I_2, \ldots\}$ with $|f(I_i)| \geq |f(I_{i+1})|$ for all $i$) and then $\sigma_\Lambda(f, \mathcal{I})$ becomes simply $\sum \frac{|f(I_i)|}{\lambda_i}$, which is a consequence of the equimonotonic sequences inequality [11, Thm. 368].

It is interesting that for any choice of (i)-(iv) below the quantity

$$\sup_{\mathcal{I}} \sigma_\Lambda(f, \mathcal{I})$$

yields the same value:

(i) the supremum runs over all possible families $\mathcal{I}$ of non-overlapping closed subintervals of $[a, b]$;
(ii) the supremum runs over all infinite families $\mathcal{I}$ of non-overlapping closed subintervals of $[a, b]$;
(iii) the supremum runs over all finite families $\mathcal{I}$ of non-overlapping closed subintervals of $[a, b]$;
(iv) the supremum runs over all finite families $\mathcal{I}$ of non-overlapping closed subintervals filling the whole interval $[a, b]$, that is, with $\bigcup \mathcal{I} = [a, b]$.

The choice of option (iv) and the particular $\Lambda$-sequence $J := (1)_{i=1}^{\infty}$ shows that in this case $\Lambda$-variation coincides with the classical Jordan variation $V(f)$ of $f$. Finite families $\mathcal{I}$ filling the whole interval $[a, b]$ correspond to partitions of the interval naturally. Thus, given a partition $\pi = (t_i)_{i=0}^{n}$ of $[a, b]$, we define a family $\mathcal{I}_\pi$ by $\mathcal{I}_\pi := \{[t_{i-1}, t_i] : i = 1, \ldots, n\}$. Then

$$\sigma_J(f, \pi) \ := \ \sigma_J(f, \mathcal{I}_\pi) \ = \ \sum_{i=1}^{n} |f(t_i) - f(t_{i-1})|$$

and

$$V(f) \ = \ V_J(f) \ = \ \sup_{\pi} \sigma_J(f, \pi) \ = \ \sup_{\pi=(t_i)_{i=0}^{n}} \sum_{i=1}^{n} |f(t_i) - f(t_{i-1})|.$$

In this way we have arrived at the classical definition of ordinary variation.

It is not difficult to show that if a function $f$ defined on a closed interval $I$ is not of bounded $\Lambda$-variation, then there is a point $x_0 \in I$ such that, for every interval $K$ containing $x_0$ in its interior (relative to $I$) $V_\Lambda(f, K) = +\infty$ [37, Lemma 1]. In case of ordinary variation this observation can be strengthened to the following form: if $f$ is not of bounded variation on $I$, then

$$\sum_{i=1}^{\infty} |f(t_i) - f(t_{i-1})| \ = \ +\infty$$

for a strictly monotone sequence $(t_i)_{i=0}^{\infty}$ of points of $I$ ( cf. [21, Prop. 4.1]). Of course, the sequence converges to a point $x_0 \in I$ and $f$ is of unbounded variation on every neighbourhood of $x_0$.

The question, whether the above implication remains valid for any $\Lambda$-sequence, leads us to the following definition [39, p.75]. A function $f$ on $I$ is said to be of bounded ordered $\Lambda$-variation ($f \in O\Lambda BV$ ) if

$$\sup \sum \frac{|f(I_n)|}{\lambda_n} \ < \ +\infty,$$

where the supremum is extended over all collections of intervals $\{I_n\}$ for which either $\sup I_n \le \inf I_{n+1}$ for all $n$ or $\sup I_{n+1} \le \inf I_n$ for all $n$. The idea of defining and investigating functions of bounded $\Lambda$-variation has its genesis in a result of Goffman and Waterman who characterized the class $GW$ of regulated functions which, for every change of variable, have an everywhere convergent Fourier series. A similar characterization for the class $UGW$ of continuous functions which, for every change of variable, have uniformly convergent Fourier series was obtained by Baernstein and Waterman. Both characterizations involve sums of the form $\sum_{n=1}^{N} \frac{f(I_n)}{n}$ and the intervals $I_n$ are ordered, i.e. either $\sup I_n \le \inf I_{n+1}$ for all $n$ or $\sup I_{n+1} \le \inf I_n$ for all $n$. This suggested that the notion of order can be added to create another variant of generalized variation.

The inclusion $\Lambda BV \subset O\Lambda BV$ is obvious, but are the sets equal? The question was raised in [39], and answered in negative for the special $\Lambda$-sequence $H = (i)_{i=1}^{\infty}$ by C.L. Belna who constructed a continuous function $f \in OHBV \setminus HBV$ [2]. Since the method of Belna was applicable to $OHBV$ only, D. Waterman put the question about the general case in his list of open questions again [41, Question 4]. It turned out that the inclusion $\Lambda BV \subset O\Lambda BV$ is proper for every proper $\Lambda$-sequence $\Lambda$ [28]. It was later rediscovered in an easily accessible paper [22].

## 2. Basic Properties

Several properties of functions of bounded $\Lambda$-variation was mentioned in the first paper where the new generalized variation was introduced [35, p.108]. In particular, every function of bounded $\Lambda$-variation is bounded and regulated, and the set $\Lambda BV$ of all functions of bounded $\Lambda$-variation is a linear space (for proofs see [18, Thms 1,3,4]). Since every function in $\Lambda BV$ has only simple discontinuities, it makes sense to say that a function $f$ has an internal saltus at a point $x_0$ of discontinuity if

$$\liminf_{x\to x_0} f(x) \;\le\; f(x_0) \;\le\; \limsup_{x\to x_0} f(x).$$

If a $\Lambda BV$ function has an internal saltus at each point of discontinuity, then its total variation is independent of its values at points of discontinuity [19, Thm. 1]. In particular, if two functions in $\Lambda BV$ agree at points of continuity and have an internal saltus at each point of discontinuity, then they have the same total $\Lambda$-variation [19, Cor.]. Further, of all functions continuous at the points of continuity of a given functions, those with internal saltus at points of discontinuity have minimal total $\Lambda$-variation [19, Thm. 2].

If $f$ is of bounded $\Lambda$-variation on $[a, b]$, then we can consider the $\Lambda$-variation function $v_{\Lambda, f}$ of $f$ on $[a, b]$ defined by

$$v_{\Lambda, f}(x) \;:=\; V_\Lambda(f, [a, x]).$$

Here we encounter the first essential difference between ordinary variation and the generalized variation. Namely, the ordinary variation is additive with respect to intervals, that is,

$$V(f,\,[a,\,b]) \;=\; V(f,\,[a,\,x]) \;+\; V(f,\,[x,\,b])$$

if $f$ is of bounded variation. On the other hand, for a general Λ-sequence, one has subadditivity only

$$V_\Lambda(f,\,[a,\,b]) \;\le\; V_\Lambda(f,\,[a,\,x]) \;+\; V_\Lambda(f,\,[x,\,b])$$

for $f \in \Lambda BV[a,\,b]$ [37, Thm. 2], and equality takes place for few special cases only.

The behaviour of Λ-variation of a function on an diminishing interval is summarized in the next theorem (cf. [37, Thm. 3] and [31, Prop. 1.3]). But first we need suitable notation: given a regulated function $f$, we denote the left-hand jump of $f$ at a point $x$ by $\Delta_- f(x) := f(x) - f(x-)$, and the right-hand jump of $f$ at the point $x$ by $\Delta^+ f(x) := f(x+) - f(x)$. We define also $\Delta f(x) := f(x+) - f(x-)$.

**Theorem 2.1.** *If $f$ is of bounded Λ-variation, then*

*(i)* $\displaystyle\lim_{x\to x_0^-} V_\Lambda\big(f,\,[x,\,x_0)\big) \;=\; 0 \;=\; \lim_{x\to x_0^+} V_\Lambda(f,\,(x_0,\,x])$ :

*(ii)*

$$\lim_{x\to x_0^-} V_\Lambda(f,\,[x,\,x_0]) \;=\; \frac{|\Delta_- f(x_0)|}{\lambda_1}$$

*and*

$$\lim_{x\to x_0^+} V_\Lambda(f,\,[x_0,\,x]) \;=\; \frac{|\Delta^+ f(x_0)|}{\lambda_1};$$

*(iii) if $f$ has an internal saltus at $x_0$, then*

$$\lim_{\substack{x\to x_0^- \\ y\to x_0^+}} V_\Lambda(f,\,[x,\,y]) \;=\; \frac{|\Delta f(x_0)|}{\lambda_1},$$

*and if $f$ has an external saltus at $x_0$, then*

$$\lim_{\substack{x\to x_0^- \\ y\to x_0^+}} V_\Lambda(f,\,[x,\,y]) \;=\; \frac{\max\{|\Delta_- f(x_0)|,|\Delta^+ f(x_0)|\}}{\lambda_1} \;+\; \frac{\min\{|\Delta_- f(x_0)|,|\Delta^+ f(x_0)|\}}{\lambda_2}.$$

Another important property of the Λ-variation function analogous to a property well-known for ordinary variation functions is the following [37, Thm. 4].

**Theorem 2.2.** *Let $f$ be a function of bounded Λ-variation. Then $v_{\Lambda,f}$ is right (left) continuous at any point of $I$ if and only if $f$ is right (left) continuous at that point.*

The theorem was later partially generalized in the following manner. Given a subset $A$ of $[a,\,b]$, the supremum of numbers $\sigma_\Lambda(f,\,\mathcal{I})$, where $\mathcal{I}$ runs over all possible families of non-overlapping closed intervals with endpoints in $A$, will be called the

$\Lambda$-variation of the function $f$ on the subset $A$ and denoted by $V_\Lambda(f, A)$. Consider now the space $\mathcal{K}$ of all closed subsets of $[a, b]$ with the standard Hausdorff metric $\delta$. Then $[a, x] \xrightarrow{\delta} [a, y]$ if and only if $x \to y$. Thus the following theorem [23, Prop. 2.9] includes a part of Thm. 2.2.

**Theorem 2.3.** *Let $f$ be a function of bounded $\Lambda$-variation. Then $f$ is continuous if and only if the function $(\mathcal{K}, \delta) \to \mathbb{R} : A \mapsto V_\Lambda(f, A)$ is continuous.*

The idea of computing $\Lambda$-variation on a subset is useful especially when the $\Lambda$-variation of $f$ on the subset is equal to the $\Lambda$-variation of $f$ on the whole interval $[a, b]$. The most important result of this kind uses the concept of a point of varying monotonicity introduced by Bruckner and Goffman. A point $x \in (a, b)$ is said to be of varying monotonicity of a function $f$ if there is no neighbourhood of $x$ on which $f$ is strictly monotone or constant. The endpoints $a$ and $b$ are said to be of varying monotonicity of $f$ if the function is non-constant on every neighbourhood of $a$ or $b$, respectively. The set of all points of varying monotonicity of a function $f$ is denoted by $K_f$. It is a closed subset of $[a, b]$ always. For any continuous function $f$ the equality $V_\Lambda(f) = V_\Lambda(f, K_f)$ holds [23, Prop. 1.1]. This method of simplifying computation of $\Lambda$-variation can be improved even further by replacing $K_f$ by any of its dense subsets [23, Cor. 1.6].

In case of ordinary variation the functions $p(x) = \frac{1}{2}(v_f(x) + f(x))$ and $n(x) = \frac{1}{2}(v_f(x) - f(x))$ yield the powerful Jordan decomposition $f = p - n$ of a function $f$ of bounded variation on $I$ into a difference of two non-decreasing functions. Since for most $\Lambda$-sequences the set $\Lambda BV$ is larger than $BV$, the decomposition is no longer valid for $\Lambda BV$-functions. Nevertheless, one can investigate a possibility of giving a characterization of $\Lambda BV$-functions similar to Jordan characterization of $BV$-functions. We can define positive and negative $\Lambda$-variations by setting

$$v^+_{\Lambda, f} := \sup\left\{\sum \frac{f(I_n)}{n} : I_n \subset [a, b],\ f(I_n) > 0\right\}$$

and

$$v^-_{\Lambda, f} := \sup\left\{\sum \frac{f(I_n)}{n} : I_n \subset [a, b],\ f(I_n) < 0\right\}$$

for a function $f$ of bounded $\Lambda$-variation on $[a, b]$. The following question remains open [37, p.44].

**Problem 2.1.** *To what extent do the positive and negative $\Lambda$-variations of a function characterize the function?*

The linear space $\Lambda BV$ can be endowed with a natural and very nice topology [37, pp. 41-42].

**Theorem 2.4.** *The functional*

$$\|f\|_\Lambda := |f(a)| + V_\Lambda(f)$$

*is a norm on the linear space $\Lambda BV$. The space $(\Lambda BV, \|\ \|_\Lambda)$ is a Banach space.*

Taking any fixed point $x_0 \in [a, b]$ instead of $a$ in the definition of the Λ-variation norm would produce an equivalent norm only. since $\Lambda BV$ functions are bounded, another possible choice of a norm on $\Lambda BV$ is

$$\|f\|_{\Lambda,\infty} \;:=\; \|f\|_\infty + V_\Lambda(f).$$

This norm is equivalent with $\|\ \|_\Lambda$ and $\Lambda BV$ under this norm $\|\ \|_{\Lambda,\infty}$ forms a commutative Banach algebra with respect to pointwise multiplication [16, Thm. 4].

Since convergence in Λ-variation norm implies uniform convergence, the space $C\Lambda BV$ of continuous functions of bounded Λ-variation is a closed subspace of $\Lambda BV$. Two more important facts, similar to those known for ordinary variation, remain true for the generalized variation.

**Theorem 2.5.** *If a sequence $(f_n)$ of $\Lambda BV$-functions converges pointwise to a function $f$, then $V_\Lambda(f) \leq \liminf V_\Lambda(f_n)$.*

Next is a generalization of Helly's extraction theorem [37, Thm. 5].

**Theorem 2.6.** *If $(f_n)$ is a sequence of $\Lambda BV$-functions with $\|f_n\|_\Lambda \leq M$ for all $n$, then there is a subsequence $(f_{n_k})$ converging pointwise to a function $f \in \Lambda BV$ with $\|f\|_\Lambda \leq M$.*

A characterization of the invariance of the class $HBV$ under composition [5] was later found to be true for all Λ-sequences [20].

**Theorem 2.7.** *$g \circ f$ is in the class $\Lambda BV$ for each $f$ of that class whose range is in the domain of $g$ if and only if $g \in Lip\,1$.*

M. Josephy found a characterization of functions $g$ such that $f \circ g \in BV$ for every $f \in BV$ [12, Thm. 3]. His proof, however, cannot be extended to the general $\Lambda BV$ case, because it depends on the Jordan decomposition of $BV$ functions.

**Problem 2.2.** *Given a Λ-sequence Λ, how can we characterize functions $g : [a, b] \to [a, b]$ such that $f \circ g \in \Lambda BV$ for every $f \in \Lambda BV$ ?*

## 3. The Structure of Regulated Functions

The fundamental characterization of inclusions between $\Lambda BV$ classes for distinct Λ-sequences Λ was found by S. Perlman and D. Waterman [19, Thm. 3].

**Theorem 3.1.**

$$\Lambda BV \subset \Gamma BV \qquad \textit{if and only if} \qquad \sum_{k=1}^{n} \frac{1}{\gamma_k} = O\left(\sum_{k=1}^{n} \frac{1}{\lambda_k}\right).$$

It follows that $\Lambda BV = \Gamma BV$ if and only if the ratios of partial sums of the Λ-sequences $\sum_1^n \frac{1}{\gamma_k} / \sum_1^n \frac{1}{\lambda_k}$ are bounded both from 0 and from $+\infty$. In that case we will say that the Λ-sequences Λ and Γ are equivalent. It has been shown in [27] that for every $\Lambda$-sequence $\Lambda$ there is an equivalent $\Lambda$-sequence $\Gamma = (\gamma_i)$ such that $\lim \frac{\gamma_{i+1}}{\gamma_i} = 1$.

The definition of a $\Lambda$-sequence implies that for every $\Lambda$-sequence $\Lambda = (\lambda_i)$ one has $\sum_{k=1}^n \frac{1}{\lambda_k} \le \frac{n}{\lambda_1}$ and thus $\sum_{k=1}^n \frac{1}{\lambda_k} = O(\sum_{k=1}^n 1)$. Thus, $BV \subseteq \Lambda BV$ for every $\Lambda$-sequence $\Lambda$ by the previous Theorem, which means that functions of bounded variation form the smallest $\Lambda BV$ class. Since $\Lambda BV = BV$ if and only if $\lim \lambda_i < +\infty$, $\Lambda$ is a proper $\Lambda$-sequence if and only if $\Lambda$-variation is a proper generalization of ordinary variation, that is, when $BV \subsetneq \Lambda BV$. On the other hand, there is no largest $\Lambda BV$ class, since for every $\Lambda$-sequence $\Lambda$ one can construct a $\Lambda$-sequence $\Gamma$ such that $\sum_{k=1}^n \frac{1}{\gamma_k} = o\left(\sum_{k=1}^n \frac{1}{\lambda_k}\right)$, which implies that $\Lambda BV \subsetneq \Gamma BV$. The following remarkable theorem is due to S. Perlman [18, Thms. 5,7,10,12].

**Theorem 3.2.** *BV is the intersection of all $\Lambda BV$ spaces, but not the intersection of any countable collection. The space of regulated functions is the union of all $\Lambda BV$ spaces, but not the union of any countable collection.*

The mutual relationship between $\Lambda BV$ classes is rather chaotic. It is easy to see that for every proper $\Lambda$-sequence $\Lambda$ there is a proper $\Lambda$-sequence $\Gamma$ such that $\Gamma BV$ is a proper subset of $\Lambda BV$, and that for every $\Lambda$-sequence $\Lambda$ there is a proper $\Lambda$-sequence $\Gamma$ such that $\Gamma BV$ contains $\Lambda BV$ properly. On the other hand the following fact holds.

**Theorem 3.3.** *For every proper $\Lambda$-sequence $\Lambda$ there is a proper $\Lambda$-sequence $\Gamma$ such that neither $\Lambda BV \subseteq \Gamma BV$ nor $\Gamma BV \subseteq \Lambda BV$.*

**Proof.** Let $(\lambda_i)$ be a proper $\Lambda$-sequence and let $\epsilon_p$ be a positive number.

*Procedure A.* Suppose that positive numbers $\gamma_1 \le \gamma_2 \le \ldots \le \gamma_N$ are given. The procedure A shows how we can find a positive integer $M > N$ and define numbers $\gamma_{N+1} \le \gamma_{N+2} \le \cdots \le \gamma_M$ so that $\gamma_N \le \gamma_{N+1}$ and

$$\frac{\sum_{i=1}^M \frac{1}{\gamma_i}}{\sum_{i=1}^M \frac{1}{\lambda_i}} < \epsilon_p.$$

Since $\Lambda$ is proper, we can find easily $M$ such that

$$M > N, \qquad \frac{\sum_{i=1}^N \frac{1}{\gamma_i}}{\sum_{i=1}^M \frac{1}{\lambda_i}} < \frac{\epsilon_p}{2} \qquad \text{and} \qquad \sum_{i=1}^M \frac{1}{\lambda_i} \ge 1.$$

Then we define

$$\gamma_i := \max\left\{\frac{2(M-N)}{\epsilon_p}, \lambda_i, \gamma_N\right\}$$

for $i = N+1, \ldots, M$, and obtain

$$\frac{\sum_{i=1}^M \frac{1}{\gamma_i}}{\sum_{i=1}^M \frac{1}{\lambda_i}} = \frac{\sum_{i=1}^N \frac{1}{\gamma_i}}{\sum_{i=1}^M \frac{1}{\lambda_i}} + \frac{\sum_{i=N+1}^M \frac{1}{\gamma_i}}{\sum_{i=1}^M \frac{1}{\lambda_i}} < \frac{\epsilon_p}{2} + \frac{(M-N)\frac{\epsilon_p}{2(M-N)}}{1} = \epsilon_p.$$

*Procedure B.* Suppose that positive numbers $\gamma_1 \le \gamma_2 \le \ldots \le \gamma_N$ are given. The procedure B shows how we can find a positive integer $M > N$ and define numbers $\gamma_{N+1} \le \gamma_{N+2} \le \ldots \le \gamma_M$ so that $\gamma_N \le \gamma_{N+1}$ and

$$\frac{\sum_{i=1}^{M} \frac{1}{\lambda_i}}{\sum_{i=1}^{M} \frac{1}{\gamma_i}} < \epsilon_p.$$

Since the arithmetic average of initial terms of a sequence convergent to 0 converges to 0, we have

$$\alpha_k := \frac{\sum_{i=1}^{k} \frac{1}{\lambda_i}}{\sum_{i=1}^{N} \frac{1}{\gamma_i} + (k-N)\frac{1}{\gamma_N}} = \frac{1}{\frac{\sum_{i=1}^{N} \frac{1}{\gamma_i}}{k} + \frac{k-N}{N} \cdot \frac{1}{\gamma_N}} \cdot \frac{\sum_{i=1}^{k} \frac{1}{\lambda_i}}{k},$$

and hence

$$\alpha_k \xrightarrow[k\to\infty]{} \frac{1}{0 + 1 \cdot \frac{1}{\gamma_N}} \cdot 0 = 0.$$

Thus we can find $M > N$ such that $\alpha_M < \epsilon_p$. Then we can define $\gamma_i := \gamma_N$ for $i = N+1, \ldots, M$, and the description of procedure B is complete.

Finally, starting with a proper $\Lambda$-sequence $(\lambda_i)$ and using procedures A and B interchangingly with $\epsilon_p$ tending to 0 ($p$ can be the number of a successive step of the construction), we build a $\Lambda$-sequence $(\gamma_i)$ with the desired properties. Note that the term $\lambda_i$ involved in the definition of $\gamma_i$ in procedure A guarantees that the $\Lambda$-sequence $(\gamma_i)$ is proper. □

S. Perlman and D. Waterman have shown, in the course of the proof of Thm. 3 from [19], that if there is a constant $C > 0$ such that

$$\sum_{i=1}^{n} \frac{1}{\lambda_i} < C \sum_{i=1}^{n} \frac{1}{\gamma_i} \qquad \text{for all } n,$$

then, given any non-increasing sequence $(a_i)$ of non-negative numbers,

$$\sum_{i=1}^{n} \frac{a_i}{\lambda_i} \le C \sum_{i=1}^{n} \frac{a_i}{\gamma_i}.$$

It follows that $\|f\|_\Lambda \le (C+1)\|f\|_\Gamma$ for every $f \in \Gamma BV$. Thus if $\Gamma BV \subsetneq \Lambda BV$, then the inclusion $A : \Gamma BV \to \Lambda BV$ is a continuous linear operator between Banach spaces and hence $A(\Gamma BV) = \Gamma BV$ is $\sigma$-strongly porous in $\Lambda BV$ by the Banach-Steinhaus-Olevskii Theorem [17], since we have assumed $\Gamma BV \ne \Lambda BV$ (cf. [7, Thm.3]). In particular, since no Banach space is $\sigma$-strongly porous in itself, we obtain the following equivalent conditions [6, Thm. 2].

**Theorem 3.4.** *Let* $\Lambda = (\lambda_i)$ *and* $\Lambda^{(n)} = (\lambda_i^n)_{i\in\mathbb{N}}$, $n = 1, 2, \ldots,$ *be* $\Lambda$*-sequences. Then the following statements are equivalent:*

*(i)* $\Lambda BV \subset \bigcup_{n=1}^{\infty} \Lambda^{(n)}BV$;

*(ii)* $\Lambda BV \subset \Lambda^{(N)}BV$ *for some positive integer* $N$;

*(iii)* *there is an index* $N$ *such that*

$$\sum_{i=1}^{k} \frac{1}{\lambda_i^N} = O\left(\sum_{i=1}^{k} \frac{1}{\lambda_i}\right) \qquad \text{as } k \to \infty.$$

We can now conclude as a corollary that a union of two $\Lambda BV$ classes is a $\Lambda BV$ class if and only if one of the classes contains the other. However, for any countable family of $\Lambda BV$ classes there is a $\Lambda BV$ class that contains all classes belonging to the family, which follows from Thm. 3.1, since for every countable family of divergent series there is always a divergent series diverging less rapidly than every one of the members of the family [13, p.302] On the other hand, an intersection of a finite family of $\Lambda BV$ spaces always is a $\Lambda BV$ space [18, Thm. 6] (cf. also [25, Lemma 16]). However, intersections of countably infinite families of $\Lambda BV$ spaces are curious. Namely, we have the following fact [6, Thm. 1] (see also [25, Thm. 10]).

**Theorem 3.5.** *Let* $\Lambda = (\lambda_i)$ *and* $\Lambda^{(n)} = (\lambda_i^n)_{i\in\mathbb{N}}$, $n = 1, 2, \ldots,$ *be* $\Lambda$*-sequences. Then the following statements are equivalent:*

*(i)* $\bigcap_{n=1}^{\infty} \Lambda^{(n)}BV \subset \Lambda BV$;

*(ii)* $\bigcap_{n=1}^{N} \Lambda^{(n)}BV \subset \Lambda BV$ *for some positive integer* $N$;

*(iii)* *there is an index* $N$ *such that*

$$\sum_{i=1}^{k} \frac{1}{\lambda_i} = O\left(\max_{1\le n\le N} \sum_{i=1}^{k} \frac{1}{\lambda_i^n}\right) \qquad \text{as } k \to \infty.$$

There is one nice property concerning intersections of uncountable families of $\Lambda BV$ classes which generalizes a part of Perlman's result [25, Prop. 9 and Cor. 3].

**Theorem 3.6.** *For any* $\Lambda$*-sequence* $\Lambda = (\lambda_i)$

$$\Lambda BV = \bigcap_{\Lambda = o(\Gamma)} \Gamma BV,$$

*where the intersection is taken over all* $\Lambda$*-sequences* $\Gamma$ *such that* $\lambda_i = o(\gamma_i)$. *Moreover, the equality cannot be achieved by an intersection of any countable subfamily.*

Now one could wonder if there is a result similar to Thm. 3.6 involving a union, that is, if the equality $\Lambda BV = \bigcup_{\Gamma = o(\Lambda)} \Gamma BV$ holds. The answer is known, but not so simple, and we have to postpone it till later, because we lack the suitable language to describe the situation precisely.

The paper [7] contains a number of results relating $\Lambda$-variation and porosity. One of them is that $C\Lambda BV$ is strongly porous in $\Lambda BV$ [7, Thm. 4]. It is also shown that the set of all upper semi-continuous functions of bounded $\Lambda$-variation is strongly porous in $\Lambda BV$ [7, Thm.7]. Another result of similar kind is that the set of all continuous functions of bounded $\Lambda$-variation that are differentiable at least at one point is a dense and first Baire category subset of $C\Lambda BV$ for any proper $\Lambda$-sequence [24, Prop. 1 and Prop. 2].

Regulated functions equipped with the standard sup-norm, $\|f\| = \sup_{x\in I}|f(x)|$, form a Banach space. Since $\|f\| \leq (1+\lambda_1)\|f\|_\Lambda$ for any $\Lambda$-sequence $\Lambda = (\lambda_i)$ and any $f \in \Lambda BV$, the Olevskii's version of the Banach-Steinhaus Theorem tells us that $\Lambda BV$ is $\sigma$-strongly porous in the space of regulated functions, and this yields another proof that regulated functions are not the union of a countable collection of $\Lambda BV$ classes.

D. Waterman proved that $O\Lambda BV$ is a Banach space under the norm $\|f\|_\Lambda^{ord} := |f(a)| + V_\Lambda^{ord}(f)$ where $V_\Lambda^{ord}(f)$ denotes the ordered $\Lambda$-variation of $f$ [42, Prop. 2]. Since $\|f\|_\Lambda^{ord} \leq \|f\|_\Lambda$ for any $f \in \Lambda BV$, it follows from [17, Lemma] that $\Lambda BV$ is a strongly porous set in $O\Lambda BV$. In particular $\Lambda BV$ is of first Baire category in $(O\Lambda BV, \|\ \|_\Lambda^{ord})$ [42, Prop. 3].

## 4. Local $\Lambda$-Variation and Its Decomposition

We are going to alter the definition of $\Lambda$-variation slightly. Given a family $\mathcal{I}$ of non-overlapping closed subintervals of $[a, b]$ and a function $f$ defined on the interval, we set $\|\mathcal{I}\| := \max_{I\in\mathcal{I}}|I|$ and $\|\mathcal{I}\|_f := \sup_{I\in\mathcal{I}}|f(I)|$. Next, given a $\Lambda$-sequence $\Lambda$ and a positive number $\delta$, we define

$$V_{\Lambda,\delta}(f) \ := \ \sup \sigma_\Lambda(f, \mathcal{I}),$$

where the supremum is taken over all families $\mathcal{I}$ with $\|\mathcal{I}\| \leq \delta$. The value

$$W_\Lambda(f) \ := \ \lim_{\delta\to 0+} V_{\Lambda,\delta}(f)$$

will be called the Wiener $\Lambda$-variation of $f$ [23]. This does not lead to any new class of functions for $W_\Lambda(f)$ is finite if and only if $V_\Lambda(f)$ is finite. One has $W_\Lambda(f) \leq V_\Lambda(f)$ always, and the inequality may be strict which is the case, for instance, for any non-constant monotone function $f$ and any proper $\Lambda$-sequence $\Lambda$.

The functional

$$\|f\|_\Lambda^W \ := \ |f(a)| \ + \ W_\Lambda(f)$$

is a seminorm on $\Lambda BV$.

Another useful modification of the concept of $\Lambda$-variation is the following. Given a $\Lambda$-sequence $\Lambda$ and a positive number $\delta$, we set

$$V_\Lambda^\delta(f) \ := \ \sup \sigma_\Lambda(f, \mathcal{I}),$$

where the supremum is taken over all families $\mathcal{I}$ with $\|\mathcal{I}\|_f \leq \delta$, and then we define

$$V_\Lambda^0(f) \ := \ \lim_{\delta\to 0+} V_\Lambda^\delta(f).$$

Clearly, $V_\Lambda^0(f) \le V_\Lambda(f)$ for every function. If $f$ is regulated, then $V_\Lambda^0(f)$ is finite if and only if $V_\Lambda(f)$ is finite. As earlier, the functional

$$\|f\|_\Lambda^0 \ := \ |f(a)| \ + \ V_\Lambda^0(f)$$

is a seminorm on $\Lambda BV$. Exactly like for $\Lambda$-variation, the computation of Wiener $\Lambda$-variation of a continuous function requires the knowledge of points of varying monotonicity only [23, Prop. 1.3].

**Theorem 4.1.** *For any proper $\Lambda$-sequence $\Lambda$ and any continuous function $f$*

$$W_\Lambda(f) \ = \ W_\Lambda(f, K_f).$$

Next, given a regulated function $f$ on $[a, b]$, we agree the if $f$ has an internal saltus at a point $x$, then the single jump of $f$ at $x$ is $|\Delta f(x)|$, and if $f$ has an external saltus at a point $x$, then $f$ has two jumps at the point: $|\Delta_- f(x)|$ and $|\Delta^+ f(x)|$. Then we arrange all jumps of $f$ in a non-increasing order $(\eta_i^f)$. If $f$ has only finitely many discontinuities, we fill up the rest of the sequence $(\eta_i^f)$ with zeros. Clearly, $f \in \Lambda BV$ implies $\sum \frac{\eta_i^f}{\lambda_i} < +\infty$, but not conversely. However, the functional

$$|||f|||_\Lambda \ := \ \sum_{i=1}^{\infty} \frac{\eta_i^f}{\lambda_i}$$

is a seminorm on $\Lambda BV$ for every $\Lambda$-sequence $\Lambda$.

The definitions allow us to formulate an important decomposition theorem [25, Thm. 1].

**Theorem 4.2.**

$$\|f\|_\Lambda^W \ = \ \|f\|_\Lambda^0 \ + \ |||f|||_\Lambda$$

*for every $\Lambda$-sequence $\Lambda$ and any regulated function $f$.*

Now, borrowing an idea of L.C. Young, we can define

$$\Lambda BV^* \ := \ \left\{ f \in \Lambda BV : \ W_\Lambda(f) = \sum \frac{\eta_i^f}{\lambda_i} \right\}$$

(see [43, p. 261] or [8, p.90]). An easy example of a function in the class $\Lambda BV^*$ is provided by any step function. It follows from Thm. 4.2 that a function $f$ of bounded $\Lambda$-variation belongs to the class $\Lambda BV^*$ if and only if $V_\Lambda^0(f) = 0$. Moreover, $(\Lambda BV^*, \| \ \|_\Lambda)$ is a closed subspace of $\Lambda BV$.

For the ordinary variation, that is, for the $\Lambda$-sequence $J = (1)_{i=1}^\infty$, we get $V(f) = V_J(f) = W_J(f)$ for every regulated function $f$. Thus Thm. 4.2 is a generalization of the well-known fact that $V(f) = V(f_c) + V(f_s)$ where $f_s$ is the saltus function of $f$ and $f_c := f - f_s$ is the continuous part of $f$ (see [33, p. 308] or, for modern and more general treatment, Cor. 7.7, Thm. 5.5 and Thm. 5.3 of [10]). Hence, we might think of functions in $\Lambda BV^*$ as "saltus-like"functions whose Wiener $\Lambda$-variation comes from jumps only.

## 5. Continuity in Λ-Variation and Λ-Absolute Continuity

The notion of continuity in Λ-variation was introduced by D. Waterman in [36] to provide a sufficient condition for $(C, \beta)$-summability of Fourier series of a function. For any Λ-sequence $\Lambda = (\lambda_i)_{i=1}^{\infty}$ and any positive integer $m$, the sequence $\Lambda_{(m)} := (\lambda_i)_{i=m+1}^{\infty}$ is a Λ-sequence as well. A function $f$ is said to be continuous in Λ-variation ($f \in \Lambda BV_c$) if $\lim_{m\to\infty} V_{\Lambda_{(m)}}(f) = 0$. The limit exists for every $f \in \Lambda BV$, because $(V_{\Lambda_{(m)}}(f))_{m\in\mathbb{N}}$ is a non-increasing sequence of non-negative numbers. It turned out later that a number of good properties of Fourier series can be proved for functions in $\Lambda BV_c$ [1]. In 1982 S. Wang found the following description of the class $\Lambda BV_c$ [34, Thm. 1], answering D. Waterman's call for a characterization of $\Lambda BV_c$ functions [37, Problem 1].

**Theorem 5.1.** *The necessary and sufficient condition for $f \in \Lambda BV_c$ is that there is a Λ-sequence $\Gamma = (\gamma_i)$ such that $\gamma_i = o(\lambda_i)$ and $f \in \Gamma BV$, that is,*

$$\Lambda BV_c \quad = \bigcup_{\Gamma = o(\Lambda)} \Gamma BV.$$

Prop. 5 of [25] tells us that $\lim_{m\to\infty} \|f\|_{\Lambda_{(m)}} = \|f\|_{\Lambda}^{0}$ for any proper Λ-sequence Λ and any regulated function $f$. From this we deduce another characterization of continuity in Λ-variation.

**Theorem 5.2.** *Let Λ be a proper Λ-sequence and let $f$ be regulated. Then $f \in \Lambda BV_c$ if and only if $V_{\Lambda}^{0}(f) = 0$.*

Hence, by Thm. 4.2, $f$ is continuous in Λ-variation if and only if $\|f\|_{\Lambda}^{W} = |f(a)| + |||f|||_{\Lambda}$, which leads to the following corollary [25, Cor. 1].

**Theorem 5.3.** $\Lambda BV_c = \Lambda BV^*$ *for every proper Λ-sequence Λ. In particular, $\Lambda BV_c$ is a closed subspace of $(\Lambda BV, \|\ \|_{\Lambda})$.*

It follows from Thm. 5.1 that $\Gamma = o(\Lambda)$ implies the inclusion $\Gamma BV \subset \Lambda BV_c$ This can be improved slightly to the form: $\sum_{i=1}^{n} \frac{1}{\lambda_i} = o(\sum_{i=1}^{n} \frac{1}{\gamma_i})$ implies that $\Gamma BV \subset \Lambda BV_c$ [25, Prop. 14], but even the condition is not necessary as it will follow from Thm. 6.1.

**Problem 5.1.** *Characterize the inclusion $\Gamma BV \subset \Lambda BV_c$ ?*

It seems that the mystery of good properties of Fourier series of functions in $\Lambda BV_c$ can be explained by the next important feature of functions continuous in Λ-variation [25, Thm. 3].

**Theorem 5.4.** *$\Lambda BV_c$ is the $\|\ \|_{\Lambda}$-closure of the set of all step functions.*

Hence $\Lambda BV_c$ has in the world of functions of bounded Λ-variation the same place as the the Wiener class $\mathcal{W}_p^*$ in the world of functions of bounded $p$-variation [15, Thms 18 and 19].

Functions continuous in $\Lambda$-variation can be quite bad since for every proper $\Lambda$-sequence $\Lambda$ there is a $\Lambda BV_c$ function which does not admit a finite approximate derivative at any point [29, Thm. 1].

Now we want to examine properties of continuous functions that are continuous in $\Lambda$-variation, $C\Lambda BV_c$. Two fundamental characterizations of functions in this class are given by the next theorem [25, Thm. 2 and Prop.6].

**Theorem 5.5.** *For every proper $\Lambda$-sequence $\Lambda = (\lambda_i)$ the following statements are equivalent:*

*(i)* $f \in C\Lambda BV_c$ ;
*(ii)* $W_\Lambda(f) = 0$ ;
*(iii)* *for every* $\epsilon > 0$ *there is a* $\delta > 0$ *such that*

$$\sum_{i=1}^{n} \frac{|I_i|}{\lambda_i} < \delta \qquad \textit{implies} \qquad \sum_{i=1}^{n} \frac{|f(I_i)|}{\lambda_i} < \epsilon$$

*for every finite family* $\{I_i\}$ *of non-overlapping closed subintervals of* $[a, b]$.

Since, for the special $\Lambda$-sequence $J = (1)_{i=1}^{\infty}$, the condition (iii) becomes the definition of classical absolute continuity, we will say that $C\Lambda BV_c$ functions are $\Lambda$-absolutely continuous and write $f \in \Lambda AC$ equivalently with $f \in C\Lambda BV_c$. Note that this definition of a $\Lambda$-absolutely continuous function yields a distinct class of functions from the class investigated by I. Lahiri in [14]. Of course, $(\Lambda AC, \|\ \|_\Lambda)$ is a closed and separable subspace of $\Lambda BV$ for it is an intersection of a closed subspace $C\Lambda BV$ and a separable and closed subspace $\Lambda BV_c$. It has been shown in the course of the proof of Prop. 3.6 from [23] that the set of all rational polygonal functions on $[0, 1]$ is $\|\ \|_\Lambda$-dense in $\Lambda AC[a, b]$, where by a rational polygonal function we mean a function $f : [0, 1] \to \mathbb{R}$ such that there is a partition $(t_i)_{i=0}^{n}$ with all points rational, and moreover, $f(t_i)$ is rational, for all $i$ and $f$ is linear on each subinterval $[t_{i-1}, t_i]$, $i = 1, \ldots, n$. The classical Plessner characterization of absolute continuity in terms of translates of a function remains true for the generalized absolute continuity which yields one more argument in support of this choice of name for functions in $C\Lambda BV_c$ [25, Thm. 4].

**Theorem 5.6.** *Let $\Lambda$ be a $\Lambda$-sequence and let $f$ be a measurable function. Then $f$ is $\Lambda$-absolutely continuous if and only if $V_\Lambda(f_h - f, [a, b-h]) \to 0$ as $h \to 0^+$ where $f_h(x) := f(x+h)$.*

Also, a characterization of compactness in $\Lambda AC$ is known [23, Prop. 3.5].

**Theorem 5.7.** *Let $\Lambda$ be a proper $\Lambda$-sequence. A set $G \subset \Lambda AC$ is precompact if and only if $\sup_{f\in G} \|f\|_\Lambda < +\infty$ and $\lim_{\delta\to 0^+} \sup_{f\in G} V_{\Lambda,\delta}(f) = 0$.*

A number of facts about inclusions between spaces $\Lambda BV$, $\Lambda BV_c$, $\Lambda AC$, analogous to those from the second section of this paper can be found in [25]. In particular, the union of all possible $\Lambda AC$ classes is the family of all continuous functions. Therefore, every continuous function $f : [a, b] \to \mathbb{R}$ can be made Λ-absolutely continuous by a suitable choice of the Λ-sequence Λ.

## 6. Relationship Between $\Lambda BV$ and $\Lambda BV_c$

As soon at it had been discovered that $\Lambda BV_c$ functions admit much better estimates of their Fourier coefficients ( [34], [29]), the question posed in [39] whether the obvious inclusion $\Lambda BV_c \subseteq \Lambda BV$ is proper or not, became quite important. The first result in this direction was presented by R. Fleissner and J. Foran who constructed a marvelous example of a special Λ-sequence Λ for which $\Lambda BV_c$ was strictly smaller than $\Lambda BV$ [9]. Another example of proper inclusion $\Lambda BV_c \subsetneq \Lambda BV$ was presented by G. Shao in [32] who first proved that [32, Thm. 3]

$$v(n, f) \quad = \quad o\left(\frac{n}{\sum_{i=1}^{n} \frac{1}{\lambda_i}}\right) \tag{1}$$

for every function $f$ continuous in Λ-variation, where $v(n, f)$ denotes the Chanturiya's modulus of variation of a function, and then constructed a Λ-sequence Λ such that

$$v(n, f) \quad \neq \quad o\left(\frac{n}{\sum_{i=1}^{n} \frac{1}{\lambda_i}}\right)$$

for a function $f \in \Lambda BV$ [32, Thm. 4]. Analogous considerations (but using an estimate on the order of magnitude of Fourier coefficients instead on of modulus of variation) lead A.I. Sablin to the conclusion that if

$$\frac{\sum_{i=1}^{\lfloor n\alpha \rfloor + 1} \frac{1}{\lambda_i}}{\sum_{i=1}^{n} \frac{1}{\lambda_i}} \quad \to \quad \alpha \tag{2}$$

for some $\alpha \in (0, 1)$, then $\Lambda BV_c$ is a proper subset of $\Lambda BV$ [29, Thm. 6]. We know today [26, Prop. 2.3] that Sablin's condition (2) is equivalent to

$$\frac{\sum_{i=1}^{2n} \frac{1}{\lambda_i}}{\sum_{i=1}^{n} \frac{1}{\lambda_i}} \quad \to \quad 2.$$

Shao observed also that if $\sum_{i=1}^{2n} \frac{1}{\lambda_i} / \sum_{i=1}^{n} \frac{1}{\lambda_i} \to 1$ as $n \to \infty$, then (1) holds for every function $f \in \Lambda BV$ [32, Thm. 5]. He proved further that if Λ is a convex positive sequence which is non-decreasing and tending to infinity, then $\Lambda BV = \Lambda BV_c$. In particular, $HBV = HBV_c$ [32, Thm. 6]. Shao also remarked that such convex Λ-sequences must satisfy $\sum_{i=1}^{2n} \frac{1}{\lambda_i} / \sum_{i=1}^{n} \frac{1}{\lambda_i} \to 1$.

Two years later A.I. Sablin announced a strong result concerning the relationship between classes $\Lambda BV$ and $\Lambda BV_c$. Namely, if the limit $\lim_{n\to\infty} \frac{\lambda_n}{\lambda_{2n}}$ exists, then in order for classes $\Lambda BV$ and $\Lambda BV_c$ to coincide, it is necessary and sufficient that the limit be less than 1 [30, Thm. 1].

A complete characterization of the equality $\Lambda BV = \Lambda BV_c$ has been presented recently [26]. Given a $\Lambda$-sequence $\Lambda = (\lambda_i)$, the Shao-Sablin index of $\Lambda$ is defined by

$$S_\Lambda \;:=\; \limsup_{n\to\infty} \frac{\sum_{i=1}^{2n} \frac{1}{\lambda_i}}{\sum_{i=1}^{n} \frac{1}{\lambda_i}}.$$

Since $\Lambda$ is a $\Lambda$-sequence, one has $S_\Lambda \in [1, 2]$ always. Moreover, if $\lim \frac{\lambda_n}{\lambda_{2n}} = \alpha$, then $S_\Lambda = 2\alpha$.

**Problem 6.1.** *Is it true that equivalent $\Lambda$-sequences have the same Shao-Sablin index?*

We conjecture that the answer is positive. The fact that if two $\Lambda$-sequences $\Lambda$ and $\Gamma$ are equivalent, then their Shao-Sablin indices are either both equal to 2 or both less than 2 [26, Prop. 2.7], can be considered to be in support of this hypothesis.

The Shao-Sablin index of a $\Lambda$-sequence is the appropriate tool for characterizing the equality $\Lambda BV = \Lambda BV_c$ as the following theorem shows [26, Thm. 3.1].

**Theorem 6.1.** *The following statements are equivalent for every proper $\Lambda$-sequence $\Lambda$:*

*(i)* *the space* $(C\Lambda BV, \|:\|_\Lambda)$ *is separable;*
*(ii)* $C\Lambda BV = C\Lambda BV_c$ *;*
*(iii)* $\Lambda BV = \Lambda BV_c$ *;*
*(iv)* $S_\Lambda < 2$.

This theorem enables us to solve another old problem [38, Problem 3] concerning the relation between $HBV$ and the Garsia-Sawyer class [26, Prop. 3.2 and Cor. 3.3].

**Theorem 6.2.** *$BV$ is a dense subspace of $(\Lambda BV, \| \, \|_\Lambda)$ if only $S_\Lambda < 2$. In particular, the Garsia-Sawyer class $GS$ is dense in $HBV$.*

The Perlman's observation that every fixed regulated function is of bounded $\Lambda$-variation for a suitably chosen $\Lambda$-sequence $\Lambda$ remains true with the additional requirement that $S_\Lambda < 2$ [26, Prop. 3.6].

**Problem 6.2.** *If $S_\Lambda = 2$, are there functions $f \in C\Lambda BV \setminus C\Lambda BV_c$ such that $W_\Lambda(f) = V_\Lambda(f)$? Such functions could be called to be $\Lambda$-singular ( $f \in S\Lambda BV$). It is not difficult to see that for a $\Lambda$-singular function $f$ the $\Lambda$-variation function $V_{\Lambda, f}$ would be an additive interval function.*

*If $\Lambda$-singular function exist, one could ask a quite bold question: for every function $f \in \Lambda BV$ (with $S_\Lambda = 2$), do there exist functions $f_{ac} \in \Lambda AC$, $f_j \in \Lambda BV_c$ and $f_s \in S\Lambda BV$ such that*

$$f \;=\; f_{ac} + f_j + f_s$$

*and*

$$V_\Lambda(f) \;=\; V_\Lambda(f_{ac}) + W_\Lambda(f_j) + V_\Lambda(f_s) \;?$$

*A positive answer to this question could be considered a generalization of the Study-Lebesgue decomposition of a function of bounded ordinary variation: for every function $f \in BV$ there are an absolutely continuous function $f_{ac}$, a saltus function $f_j$ and a singular function $f_s$ such that $f = f_{ac} + f_j + f_s$ and $V(f) = V(f_{ac}) + V(f_j) + V(f_s)$ [10], [33], [4].*

*An analogous question in the case $S_\Lambda < 2$ reads as follows: if $f \in \Lambda BV$ with $S_\Lambda < 2$, are there functions $f_{ac} \in \Lambda AC = C\Lambda BV$ and $f_j \in \Lambda BV_c = \Lambda BV^*$ such that*

$$f = f_{ac} + f_j$$

*and*

$$V_\Lambda(f) = V_\Lambda(f_{ac}) + W_\Lambda(f_j) \ ?$$

## References


[1] M. Avdispahič, *Concepts of generalized bounded variation and the theory of Fourier series*, Internat. J. Math. Math. Sci. 9(1986) 223-244

[2] C.L. Belna, *On ordered harmonic bounded variation*, Proc. Amer. Math. Soc. 80(1980) 441-444

[3] E.I. Berezhnoi, *Spaces of functions of generalized bounded variation. II. Problems of the uniform convergence of Fourier series (Russian)*, Sibirsk. Math. Zh. 42(2001) 515-532; transl. in Siberian Math. J. 42(2001) 435-449

[4] F.S. Cater, *Most monotone functions are not singular*, Amer. Math. Monthly 89(1982) 466-469

[5] M. Chaika, D. Waterman, *On the invariance of certain classes of functions under composition*, Proc. Amer. Math. Soc. 43(1974) 345-348

[6] J. Ciemnoczoowski, *Inclusion theorems for classes of functions of bounded $\lambda$-variation*, Funct. et Approx. 15(1986) 77-79

[7] R. Drozdowski, *On the structure of some subsets in the space of functions of bounded $\lambda$-variation*, Tatra Mt. Math. Publ. 34(2006) 19-27

[8] R.M. Dudley, R. Norvaiša, *Differentiability of six operators on nonsmooth functions and p-variation*, Lecture Notes in Math., Vol. 1703, Springer Verlag, Berlin - Heidelberg - New York 1999

[9] R. Fleissner, J. Foran, *A note on $\Lambda$-bounded variation*, Real Analysis Exch. 4(1978/79) 185-191

[10] K.M. Garg, *Relativization of some aspects of the theory of functions of bounded variation*, Dissertationes Math. 320(1992)

[11] G. Hardy, J.E. Littlewood, G. Pólya, *Inequalities*, 2nd edn., Cambridge University Press, Cambridge 1952

[12] M. Josephy, *Composing functions of bounded variation*, Proc. Amer. Math. Soc. 83(1981) 354-356

[13] K. Knopp, *Theory and application of infinite series*, Dover Publications Inc., New York 1990

[14] I. Lahiri, *$\Lambda$-absolute continuity relative to a set*, Bull. Inst. Math. Acad. Sinica 27(1999) 147-162

[15] E.R. Love, L.C. Young, *On fractional integration by parts*, Proc. London Math. Soc. 44(1938) 1-35

[16] L. Maligranda, W. Orlicz, *On some properties of functions of generalized variation*, Mh. Math. 104(1987) 53-65

[17] V. Olevskii, *A note on the Banach-Steinhaus theorem*, Real Analysis Exch. 17(1991/92) 399-401

[18] S. Perlman, *Functions of generalized variation*, Fund. Math. 105(1980) 199-211

[19] S. Perlman, D. Waterman, *Some remarks on functions of $\Lambda$-bounded variation*, Proc. Amer. Math. Soc. 74(1979) 113-118

[20] P. Pierce, D. Waterman, *On the invariance of classes $\Phi BV$, $\Lambda BV$ under composition*, Proc. Amer. Math. Soc. 132(2003) 775-760

[21] F. Prus-Wiśniowski, *Some remarks on functions of bounded $\phi$-variation*, Comment. Math. 30(1990) 147-166

[22] F. Prus-Wiśniowski, *On ordered $\Lambda$-bounded variation*, Proc. Amer. Math. Soc. 109(1990) 375-383

[23] F. Prus-Wiśniowski, *$\lambda$-variation and Hausdorff distance*, Math. Nachr. 158(1992) 283-297

[24] F. Prus-Wiśniowski, *$\Lambda$-variation and Baire category*, Real Analysis Exch. 20(1994/95) 134-139

[25] F. Prus-Wiśniowski, *$\Lambda$-absolute continuity*, to appear in the Rocky Mt. Math. J.

[26] F. Prus-Wiśniowski, *Separability of the space of continuous functions that are continuous in $\Lambda$-variation*, J. Math. Anal. Appl. 344(2008) 274–291

[27] F. Prus-Wiśniowski, D. Waterman, *Smoothing $\Lambda$-sequences*, Real Analysis Exch. 20(1994/195) 647-650

[28] A.A. Saakyan, *Functions of bounded $\Lambda$-variation* (Russian, Armenian summary), Akad. Nauk. Armyan. SSR Dokl. 81(1985) 54-58

[29] A.I. Sablin, *Differential properties and Fourier coefficients of functions of $\Lambda$-bounded variation*, Analysis Math. 11(1985) 331-345

[30] A.I. Sablin, *$\Lambda$-variation and Fourier series*, Soviet Math. (Iz. VUZ) 31(1987) 87-90

[31] A.I. Sablin, *Functions of bounded $\Lambda$-variation and Fourier series* (in Russian), Kandidat Thesis, Moscow State University, Moscov 1987

[32] G. Shao, *A note on the functions of $\Lambda$-bounded variation*, Chinese Ann. Math. Ser. A 6(1985) 311-316

[33] E. Study, *Über eine besondere Classe von Functionen einer reellen Veränderlichen*, Math. Ann. 47(1896) 298-316

[34] S. Wang, *Some properties of functions of $\Lambda$-bounded variation*, Scientia Sinica Ser. A 25(1982) 149-160

[35] D. Waterman, *On convergence of Fourier series of functions of generalized bounded variation*, Studia Math. 44(1972) 107-117

[36] D. Waterman, *On the summability of Fourier series of functions of $\Lambda$-bounded variation*, Studia Math. 55(1976) 97-109

[37] D. Waterman, *On $\Lambda$-bounded variation*, Studia Math. 57(1976) 33-45

[38] D. Waterman, *Bounded variation and Fourier series*, Real Analysis Exch. 3(1977-78) 61-85

[39] D. Waterman, *$\Lambda$-bounded variation: recent results and unsolved problems*, Real Analysis Exch. 4(1978/79) 69-75

[40] D. Waterman, *Fourier series of functions of $\Lambda$-bounded variation*, Proc. Amer. Math. Soc. 74(1979) 148-150

[41] D. Waterman, *Generalized bounded variation – recent results and open questions*, Real Analysis Exch. 5(1979/80) 69-75

[42] D. Waterman, *On the note of C.L. Belna*, Proc. Amer. Math. Soc. 804(1980) 445-447

[43] L.C. Young, *An inequality of the Hölder type, connected with Stieltjes integration*, Acta Math. 67(1936) 251-282

# AUTHOR INDEX